GOING BIG

GOING BIG

A SCIENTIST'S GUIDE TO LARGE PROJECTS AND COLLABORATIONS

CHRISTOPHER W. STUBBS

The MIT Press
Cambridge, Massachusetts
London, England

The open access edition of this book was made possible by generous funding and support from the author.

The MIT Press would like to thank the anonymous peer reviewers who provided comments on drafts of this book. The generous work of academic experts is essential for establishing the authority and quality of our publications. We acknowledge with gratitude the contributions of these otherwise uncredited readers.

This book was set in Bembo Book MT Pro by New Best-set Typesetters Ltd. Printed and bound in the United States of America.

Library of Congress Cataloging-in-Publication Data

Names: Stubbs, Christopher W., author.
Title: Going big : a scientist's guide to large projects and collaborations / Christopher W. Stubbs.
Description: Cambridge, Massachusetts : The MIT Press, [2024] | Includes bibliographical references and index.
Identifiers: LCCN 2023029164 (print) | LCCN 2023029165 (ebook) | ISBN 9780262547963 | ISBN 9780262378499 (epub) | ISBN 9780262378482 (pdf)
Subjects: LCSH: Research—Management. | Group work in research. | Research teams.
Classification: LCC Q180.55.M3 S78 2024 (print) | LCC Q180.55.M3 (ebook) | DDC 507.2—dc23/eng/20231016
LC record available at https://lccn.loc.gov/2023029164
LC ebook record available at https://lccn.loc.gov/2023029165

10 9 8 7 6 5 4 3 2 1

CONTENTS

PREFACE

A letter attributed to the Duke of Wellington, allegedly (Norwich 2003; Willman 2014) dispatched from Spain in August 1812:

Gentlemen

Whilst marching from Portugal to a position which commands the approach to Madrid and the French forces, my officers have been complying diligently with your requests which have been sent by H.M ship from London to Lisbon and thence by dispatch to our headquarters. We have enumerated our saddles, bridles, tents and tent poles, and all manner of sundry items for which His Majesty's Government holds me accountable. I have dispatched reports on the character, wit and spleen of every officer. Each item and every farthing has been accounted for, with two regrettable exceptions for which I beg your indulgence.

Unfortunately the sum of one shilling and ninepence remains unaccounted for in one infantry battalion's petty cash and there has been a hideous confusion as to the number of jars of raspberry jam issued to one cavalry regiment during a sandstorm in western Spain. This reprehensible carelessness may be related to the pressure of circumstance, since we are at war with France, a fact which may come as a bit of a surprise to you gentlemen in Whitehall.

This brings me to my present purpose, which is to request elucidation of my instructions from His Majesty's Government so that I may better understand why I am dragging an Army across these barren plains. I construe that perforce it must be one of two alternative duties, as given below. I shall pursue either one to the best of my ability, but I cannot do both:

1. To train an army of uniformed British clerks in Spain for the benefit of the accountants and copy-boys in London, or, perchance,
2. To see to it that the forces of Napoleon are driven from Spain.

Your most obedient servant
Wellington

For centuries, humans have undertaken ambitious projects that have required extraordinary investments of time, labor, passion, and funds. The pyramids, road networks, palaces, and religious structures of ancient times are all examples of major projects. We can presume that each of these had their own version of project management.

Many branches of modern science increasingly involve ambitious projects that are simply too large and too complex to be fully understood by any single individual. Their success requires coordination of efforts, crafting a staged schedule, and optimizing the allocation of resources. This book is an attempt to provide a resource for scientists who are ramping up the scale at which they engage in the scientific enterprise. There are two aspects to this, both addressed in this book:

1. technical project management, for the construction of large facilities, instruments, and large-scale software development, and
2. working in a large and structured science collaboration.

The Wellington quote above might capture the sentiment some scientists harbor about the mechanics of technical project management. But as the scale of a project increases, as measured in money, complexity, and number of participants, establishing a management framework is essential to a successful outcome.

This book is meant to serve as a pragmatic introductory roadmap for junior scientists who are interested in becoming more familiar with the vocabulary, methods, and practice of contemporary technical management for large scientific projects and collaborations, and as a reference for more senior scientists who are contemplating a leadership role in a major project, or who find themselves appointed to a review of the kind described below. This is the book I wish someone had handed to me when I first joined a big project.

Many of the examples are drawn from big-science projects, because that's the regime where the impact of project management is most apparent. But many of the lessons learned at the large scale can be applied to the scientific enterprise at all scales. We assume the reader is familiar with the preparation and submission of research proposals at the research-group scale. Our focus here is on the planning and execution of large collaborative projects.

My own journey down the big-project road started in graduate school with tabletop experimental physics (Adelberger et al. 1991), working in a small tight-knit team of a half-dozen scientists. That was followed by a scale-up to projects that involved tens of coauthors (e.g., Riess et al. 1998;

Alcock et al. 2000), to my eventually serving as the inaugural project scientist on the Rubin Observatory, a major ground-based astrophysics project (Ivezić et al. 2019). In parallel, I've taken on a succession of academic administrative positions, culminating in serving as Dean of Science at Harvard University. My perspective has been broadened and informed by decades of participation in JASON, a group of predominantly academic scientists who provide advice to the U.S. federal government on a wide range of technical topics. The perspectives presented here draw upon my cumulative experience and are admittedly slanted toward the U.S. scientific ecosystem. While the funding schemes and government involvement can differ in other countries, the basic principles of sound project management and effective collaborations know no borders.

Scientists involved in projects at any scale could profitably draw upon the various methods and tools of project management. For projects with total costs below half a million dollars, these tools are useful. For projects with costs in the tens of millions and above they are essential.

I am indebted to the many individuals who educated me in the world of technical project management and large-scale science collaborations. I'm particularly grateful to my colleagues Chuck Claver, Željko Ivezić, Steve Kahn, Victor Krabbendam, Robert Lupton, Vincent Riot, Steve Ritz, Don Sweeney, Kathy Turner, Tony Tyson, Kurt Vetter, Bill Wahl, and Chris Walters for their partnership on the LSST/Vera C. Rubin Observatory project. Special thanks go to Nadine Kurita for her patient tutelage. My colleagues in the Science Division office at Harvard, especially Sarah Elwell, Sharalee Field, Zoe Fonseca-Kelly, and Russ Porter, have been outstanding partners and teachers regarding the management of the science enterprise. I want to take this opportunity to recognize and salute our hard-working colleagues in the federal agencies who both support and oversee U.S. science at all scales. I'm thankful to my friends and colleagues in the JASON group, from whom I continue to learn a great deal. My colleagues Natasha Abrams, David Andrade, Sasha Brownsberger, Dillon Brout, Brodi Elwood, Peter Fisher, Eske Pedersen, John Tonry, and Elana Urbach cheered on the completion of this book. I am grateful to Nigel Poole for both encouragement and for his insightful comments. I also thank the MIT Press team who have helped make this possible: Haley Biermann, Kathleen Caruso, Andrew Clark, Emily Gutheinz, Jermey Matthews, and Mary Reilly.

My children, Rebecca and Alexander Stubbs, have provided valuable counsel and advice, for which I am very thankful, as well as sharing the perspective of a younger generation. I am grateful to my wife, Carrington Gregory, for editing assistance but more importantly for her lifelong support and encouragement, and for all she has taught me about nourishing personal relationships.

My final word of thanks goes to my Harvard colleague, Professor Jeff Lichtman. This book stems from notes made on a napkin during an enjoyable dinner conversation. Thanks for pushing me in this direction, Jeff. Sorry it took a while.

AN INTRODUCTION TO BIG PROJECTS AND LARGE TEAMS

The goal of this book is to accelerate the transition for scientists who elect to join large projects and/or big collaborations. For scientists engaged in major facilities construction or upgrades, attaining a familiarity with technical project management is essential to being an effective member of the team. A closely related topic is working and navigating within a large science collaboration, whether directly related to a facility or not.

Doing science at large scales inevitably requires imposing additional structure, both to manage the technical project and to wrangle and organize the scientific efforts of a large team. While this often leads to a sense of being embedded in a cumbersome and inefficient system, it's better than the uncoordinated chaos that is the only real alternative.

Technical project management is hard, and for big projects the stakes are high. Beyond a certain scale, project management is an essential ingredient for success. But having a project management structure in place certainly does not ensure it. NASA has undertaken a succession of flagship science projects, often called the Great Observatories. This list includes venerated and highly productive telescopes currently in orbit, including the Chandra X-ray Observatory, the optical/infrared Hubble Space Telescope, and the infrared-optimized James Webb Space Telescope (JWST). How many of NASA's Great Observatory projects came in on-time and on-budget? Zero.

The flagship NASA project that was lofted into orbit in late 2021 is the James Webb Space Telescope. This is an ambitious and well-motivated project, with a segmented mirror made of Beryllium that unfolded on orbit. The telescope is performing well and appears to be a major scientific success. But the completion of JWST has been challenging. Figure 1.1 shows an artist's rendition of how the anticipated launch date for JWST slipped over time. The cost growth has been equally daunting, increasing from initial estimates of $1 billion to the final project total at launch closer to $10 billion (Billings 2010; U.S. GAO 2021). That's an order of magnitude of

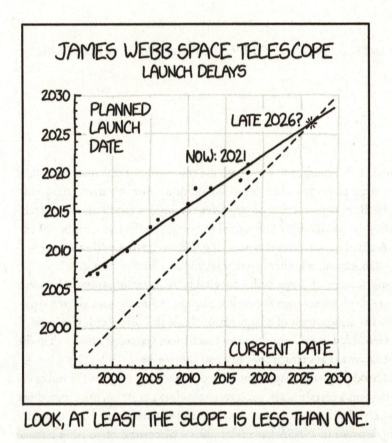

Figure 1.1
Launch date slippage for the JWST, prior to its actual launch in December 2021. The original launch date was 2007. Schedule slippage costs money. The initial $1 billion cost estimate (Billings 2010) for JWST escalated to a final project cost total closer to (U.S. GAO 2021) $10 billion. Graphic from "JWST Delays" (Munroe, n.d.).

cost growth in a project that was undertaken by an agency that is seen as a leader in technical project management. This didn't go unnoticed during the JWST construction phase. NASA made repeated trips to Capitol Hill to request augmentations to the JWST budget. The cost overruns on JWST had a severe adverse impact on smaller missions. But in the end, the JWST was completed and launched, despite the cost overruns and schedule delays.

Other projects have not been so lucky. Not all big-science projects are considered as too-big-to-fail by the elected representatives of the citizens that fund science in the United States. The Superconducting Super Collider, a particle physics experimental facility, was terminated after billions of dollars were spent, much to the chagrin of the U.S. particle physics community. While the economic and geopolitical context of the times were contributors to this decision, project management concerns, cost growth, and opposition from other subfields in physics were major factors in the cancellation decision (Riordan, Hoddeson, and Kolb 2015).

These sobering examples are not presented to cast aspersions or blame on the well-meaning people engaged in them, but rather to demonstrate just how difficult it is to first estimate and then execute an ambitious technical task that has never been done before. This stuff is hard, really hard.

On a more optimistic note, there *are* examples of major scientific projects that have come in on-time and on-budget. The Human Genome Project, undertaken by the National Institutes of Health (NIH) and the Department of Energy (DOE), was completed ahead of schedule and for less cost than initially anticipated (Olson 1993).

If you're a scientist, why should you care about the trappings of technical project management? Can't we just leave the execution of a visionary scientific initiative to professional engineers and technical administrators? I submit two categories of reasons why you should care:

1. Self-interest reasons:
   - The scientists who are most deeply engaged with the implementation of these projects are those best equipped to make knowledgeable use of their output. Understanding the nuances of the system puts you in the know.
   - Engaging with large projects at an early career stage helps you gain experience and proficiency in scientific leadership and in working with teams of people from diverse backgrounds to accomplish a shared goal.
   - Experience with real-world technical project management will give you a framework and discipline that can be usefully applied to other aspects of your life, scientific and otherwise. You'll likely write more compelling proposals, and better understand how to plan and execute projects at all scales.
   - In many cases the most challenging aspects of a scientific project are nontechnical. Dealing with the egos, differences in priorities,

personality quirks, and different talents is difficult. Gaining experience in this domain, both through your own interactions and by mindfully watching others contend with these challenges, will prove useful as you navigate this landscape.

- Evidence of effective leadership is a critical element for the career progression of scientists, both inside and outside of academia. Playing a central and successful leadership role in a large project provides clear evidence of your growth trajectory.

2. More altruistic reasons:
   - Paying it forward. Most of us have benefited from the tools, software, methods, and facilities that were built by our predecessors. We owe it to the next generation to leave a similar legacy.
   - Being part of the village . . . a blend of skills, personalities, backgrounds, and talents come together on these projects, and you can both contribute to and benefit from that breadth of community.
   - Many of these projects reach a crisis stage where a reduction in scope is needed to keep the project within budget. Finding the sweet spot between cost, performance, and schedule requires identifying the core science goals and letting go of others. This requires having the scientists at the table for those deliberations. Be there.

Spending time, effort, and energy working on a large project incurs an opportunity cost. Writing code displaces writing scientific papers. The development and refinement of instrumentation and methods for a system-of-the-future takes time away from the analysis of data from current-generation experiments. Different scientific subfields ascribe widely variable amount of "credit" to career investment in major projects. Experimental particle physics has long relied on a strong participation by practicing scientists in developing the next generation of accelerators, detectors, and analysis software, and these contributions are afforded considerable weight in career advancement decisions. Other fields have cultures that differentiate and discriminate (in the bad sense) between mere "instrument builders" and more highly prized scientists that use those tools.

One way scientists strike a balance between the near term and the long term is to split their effort between current-generation and next-generation projects. This is also a common approach for training graduate students in fields where the life cycle time for projects exceeds (or greatly exceeds) the typical time spent in graduate school.

Any specific tools and software are mentioned here mainly to illustrate concepts and principles and are not meant as endorsements. The software used for project management can range from spreadsheets and stand-alone programs on your laptop to shared-editing, cloud-based spreadsheets and scheduling tools to enterprise-level integrated programs. Using the latter can require considerable expertise, but if you're a computer-savvy scientist you can become familiar with basic project management principles over a weekend or two. Becoming proficient at technical project management takes longer. Professionals can spend their full careers at this, and there is a Project Management Certification process for these individuals.

How scientists organize themselves into collaborations is the other main topic considered in this book. While there are certainly instances where single individuals working in isolation can produce both important and interesting scientific results, in many fields there is a steady trend toward increasingly larger teams and longer authorship lists (Wuchty, Jones, and Uzzi 2007; Schulman et al. 1997).

Each of us faces a choice about the way or ways we carry out the scientific enterprise. This book should not be construed as an implicit endorsement of big-science over smaller-scale endeavors. The international scientific ecosystem thrives on projects being undertaken at a diversity of scales, ranging from a single person to the strong tradition of few-person research groups to larger collaborations to big-budget projects and facilities. Each of these scales has an important role to play, and is suited to different kinds of questions being addressed with distinct scientific sociologies.

Participating in a large collaboration does increase the complexity of doing science. There are more people drawn from a wider variety of disciplines than one normally encounters in smaller research groups, with a corresponding spectrum of values, priorities, and subfield sociologies. There are more opportunities for friction and misunderstandings. Character traits, both good and bad, that are rare in the broader population have a larger statistical chance of being present.

One very interesting but as yet unresolved question is the extent to which the rapid evolution in generative artificial intelligence (AI) will impact the planning, selection, and execution of large projects, and the nature of large science collaborations. This is a rapidly changing arena, in terms of technology, its applications, and surrounding ethics. While it is certainly tempting to turn to ChatGPT and its successors for shortcuts, be sure to keep abreast of norms and expectations. How does your funding agency feel about AI-generated proposals and reports? How do your collaborators and coauthors feel about AI-generated text for science papers? Be sure you know.

The book has three main segments. Chapters 2, 3, and 4 present a basic introduction to the fundamentals of technical project management. Chapter 5 covers the important topic of program reviews, from the perspective of both presenters and reviewers. Chapters 6 discusses various ways in which large science collaborations can be organized. Chapter 7 provides advice on how an individual scientist can be effective as a member of such a collaboration, while chapter 8 provides more specific advice in that regard for junior scientists. Chapter 9 presents some closing thoughts.

OUR CHALLENGE: BALANCING COST, RISK, AND PERFORMANCE

Faster, better, cheaper . . . pick any two.

The challenges of successfully carrying out a major project mirror what we experience in smaller ones, but with bigger budgets, more people, more time, and higher stakes. We try to do the best possible job we can with the available resources, on a reasonable timetable, anticipating risks and pitfalls. If we hit a snag then we either de-scope, find a work-around, secure additional resources to overcome the issue, or declare partial success and move on to another project.

There are differences that come with the scale and expense of a big-science project, however. For one thing, the risk tolerance of the community is, in my view, lower for big projects than it is for smaller ones. I've been on peer review panels that adopted a risk-portfolio attitude when reviewing proposals for funding, and that were willing to take a chance on a few high-risk, high-payoff proposals when taken in the context of a suite of funded projects. But if the nation is going to embark on one and only one big-budget undertaking in a given field, the community typically wants assurances that it will succeed.

One way to contend with this is to specify a set of key project performance parameters that the project team has confidence in being able to meet. These are often taken as the success/failure metrics by the funding agencies and user community.

Properly selecting key technical performance parameters is a critical decision for a project. It draws upon a combination of scientific aspiration and technical and budgetary reality. The goal is to find the sweet spot between scientific performance, cost, and risk. There is no unique figure-of-merit that applies to all situations. Experience, judgment, and ambition all play a role.

2.1   PERFORMANCE AND THE TYRANNY OF TECHNICAL REQUIREMENTS

How many data acquisition channels should we use? What's the appropriate data storage capacity? What level of redundancy do we need? What's the noise budget, and what resolution and bandwidth are needed? What sensor technology is appropriate? How will we define technical success? These are the kinds of questions that confront the early stages of a project. The answers can dictate whether the project is selected for construction, and whether it is deemed a success or a failure. It is common to identify a handful of key performance parameters (KPPs) that define the high-level technical attributes of the system, as well as methods for verifying they are met. Choosing performance parameters that are too conservative runs the risk of reviewers deciding the project is not enough of a leap forward, On the other hand, setting unrealistically ambitious goals runs the opposite risk, with the project seen as unachievable for the available budget.

There is an inevitable tension that surrounds setting these key performance parameters. Since overall project success or failure will likely be determined by whether or not they are achieved, early in the project there is an incentive to be very conservative and choose KPPs that the participants are confident they can deliver. But once chosen they will dictate the mindset of the project management team. As the project matures and confidence in eventual performance grows, scientists will find resistance from managers when they push for capabilities beyond the KPPs, even if they show those extensions will keep to the project's budget and schedule.

In an ideal world, these KPPs would flow directly from lofty science objectives. In reality, establishing a project's KPPs is the outcome of an iterative process that tries to identify the sweet spot between affordability, attainability, and ambition. These discussions often start with speculative questions like "How deep into the icecap can we imagine drilling, and what science would that enable?" The outcome of that discussion can lead to a proposal that reads as if the science goals came first, and the implementation and key project parameters followed. But in my experience that's often not the case. Proposals might well start with a science case from which the technical specifications are then "derived," but in reality it's usually a back-and-forth discussion that settles into what the key individuals consider the sweet spot. I think that's a good thing, overall, but it requires a room full of people who collectively understand both the science opportunities and the technical and budgetary realities.

The key performance parameters and the more detailed technical specifications and requirements that flow from them will rule the execution of the project, as enshrined in the Work Breakdown Structure (WBS), which is the detailed list of tasks that need to be done. Any changes in the key performance parameters ripple through the entire project, with considerable cost and schedule implications if the changes impact designs or (even worse!) constructed hardware and software. Professional project managers are *very* reluctant to make changes to these high-level performance parameters once they are articulated, reviewed, and adopted by a project. Any changes extract both political and cost/schedule tolls, usually involving a difficult negotiation with external review/oversight committees and cognizant officials from funding agencies.

The natural tendency of scientists to find ways to improve upon system performance midway through the construction phase ("Hey, if we add one more channel to the data collection system, we can measure the rate of pseudo-fermions passing through the flux gate capacitor! That would be totally cool!") will be strongly resisted by project managers who are tasked with delivering a system on time and on budget. There are instances where a certain amount of "scope creep" can be accommodated within a project's budget and schedule. Successful projects manage a positive creative tension between scientists who typically push for enhanced capabilities and managers who are tasked with delivering on time and on budget.

The conclusion I'm leading you to is this: Even if you don't choose to participate in the execution of a major project in your field, if you're a potential user then pay attention to the early stages when these key performance parameters are defined. This is usually an open and participatory process, with an opportunity for community comment and feedback. These choices matter, and you'll learn a lot by thinking about them. If you are a participant in the early stages of a project when these choices are being made, it's all the more important that you both contribute to and learn from the decision-making process. You're likely to find that it's a rather political process, where the project leadership is striving to strike a balance between the more conservative elements and more radical factions.

Once the key performance parameters and the consequent requirements and specifications are adopted, there is usually a change control process that governs any changes to them. This ensures that interfaces between subsystems are well-defined and stable, and all ripple effects for any changes are properly understood and accounted for.

The desired technical performance of course has strong implications for both cost and schedule, and the relative maturity of available technology impacts the project's risk profile.

2.2   COST AND SCHEDULE CONSIDERATIONS

The scope of items that contribute to total project cost can vary, depending on the norms of the funding agency and how the accounting system takes into consideration various contributions from participating organizations. Cost estimation is an iterative process, starting with an initial round where the scientists' ambition typically collides with limits on available funding. The upper limit on funding for a project is set by funding agency policies and/or customs, by political considerations (scientific or otherwise), or by upper limits on a particular program opportunity. An initial conceptual design is evaluated, and a cost estimate is generated. If it's deemed to exceed the funding likely to be available, the team faces a choice of either pursuing additional funding channels or else scaling back on the project's scope. This is the stage where it's important to have a combination of technical and scientific expertise at the table, to consider the tradeoffs between different implementation options, and to explore the full range of design options. We'll explore budgeting in more detail in chapter 5.

Applying the concept of confidence limits (10 percent and 90 percent confidence bounds) to cost and schedule estimates is just as important in project management as it is in presenting scientific results. Ideally, as a project progresses from the conceptual phase to the design phase to the construction phase, the uncertainties in both cost and schedule diminish.

The project's schedule is very tightly linked to the budget estimate, since personnel-related expenses are often a significant fraction of the total project cost. Of particular concern is the cost impact of schedule delays. The "marching army" costs of retaining the project's workforce for a longer period than expected is a common reason projects run over budget. Not surprisingly, in general projects that take longer cost more. This means it's worth seriously exploring whether bringing a less ambitious system online sooner might be better than spending a long time (and money) developing a more sophisticated version.

Scientists are justifiably notorious for underestimating the time needed to accomplish technical tasks. I've heard it said that scientists fall into two categories—those who underestimate by a factor of π and those who underestimate by a factor of π^2. I think there a few reasons for this. One is a combination of optimism and a well-intentioned desire to keep the associated costs to a minimum. Another is the tendency for scientists to allocate effort to solving the conceptual problem (e.g., figuring out what circuit would do the job) but not to the task itself (designing, fabricating, debugging, fabricating the inevitable second version, documenting, testing, and installing the circuit board). Another factor is the tendency to think "If I were to work only on this, how long would it take me?" If the answer is one month, the typical scientist would allocate a calendar-month of effort to the project task estimate. This does not take into account vacations, time spent in coordination meetings, documentation preparation, conducting the peer reviews needed for project execution, and (perhaps most importantly) the difference between arriving at a conceptual solution and a final, robust, tested, documented implementation of that solution. The reviews alone can easily impose a "tax" of 20 percent or more on efficiency.

2.3   TECHNICAL MATURITY AND RISK CONSIDERATIONS

Very few major projects entail bolting together existing hardware elements and running existing software. We're much more likely to be developing new capabilities or extending and improving on existing ones. The need to develop new capabilities entails associated technical risk.[1] Things usually don't work perfectly the first time, and prototyping leads to improved understanding and better hardware and software.

The less mature an essential technology, the more it presents a cost and/or schedule risk to a project. NASA has formalized the technical maturity status of flight hardware, with Technology Readiness Levels (TRLs) 1–9 as shown in table 2.1. For any proposed space project, the lower the TRL the more risk is associated with that project element. The TRL approach is discussed in more detail in chapter 4, and is presented here to give you a sense

Table 2.1
NASA Technology Readiness Levels (TRLs)

TRL	Description of Technology Readiness Levels
9	Actual system flight proven through successful mission operation
8	Actual system completed and flight qualified through test and demonstration (ground or space)
7	System prototype demonstration in a space environment
6	System/subsystem model or prototype demonstration in a relevant environment (ground or space)
5	Component and/or breadboard validation in relevant environment
4	Component and/or breadboard validation in a laboratory environment
3	Analytical and experimental critical function and/or characteristic proof-of-concept
2	Technology concept and/or application formulated
1	Basic principles observed and reported.

Note the emphasis on systems integration and on demonstrated performance in a space environment. The transition from TRL 4 to higher levels is considered a large step, and often prone to failure. Scientists tend to underestimate the effort and investment needed to advance TRL state.

of how technical managers structure their thinking about the maturity of subsystems, components, and software modules.

Active risk assessment, mitigation, and tracking is essential in modern technical project management. At all stages of project planning and execution, possible risks are identified and logged into a project "risk registry." This can be thought of as a spreadsheet with a row for each risk element, along with an initial assessment of the likelihood of it occurring, and its cost and schedule impacts.

The project team can "retire" a risk, by showing through analysis or early prototyping that is unlikely to occur, or "buy down" a risk through an early investment of project funds to arrive at a technical solution. One of your responsibilities as a project participant is to contemplate things that could go awry, with an assessment of the likelihood and consequences of this event, and have those concerns assessed and then entered into the risk registry. It is most helpful if you can also determine what steps the project can take to mitigate the concern you have raised.

As discussed in more detail below, one of the more important tasks of technical project management is to allocate resources (money, effort, and time) in order to reduce the overall risk to the program.

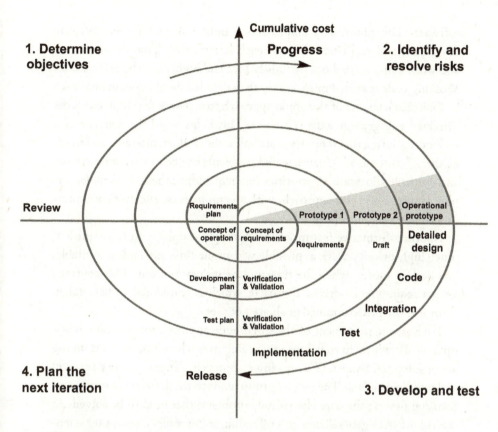

Figure 2.1
Spiral development model. One approach to software development is to just start
making what you want, as a rapid prototype, and then evolve forward through rapid
successive iterations of end-to-end solutions. This contrasts with an approach that
requires a complete and detailed design that before writing a line of code. The Agile
approach to project management embraces this concept (from Conrad Nutschan,
inspired by Boehm 1988).

2.4   TRADITIONAL VERSUS AGILE METHODS

There is something of an insurrection brewing in project management
methodology. The traditional sequential process of specifying requirements, performing detailed design, and only then actually making anything
is being challenged by proponents of "Agile" methods. The Agile approach
is derived from the spiral development method for software (Boehm 1986).
The basic idea for this is shown in figure 2.1. Rapid iterations of end-to-
end solutions are generated, in close collaboration with the users of the

software. This philosophy emphasizes a tight linkage between software developers and end users through each iteration of spiral development, and work being carried out by tightly knit and highly collaborative teams. Working code is valued more highly than detailed design documentation.

One characteristic of this spiral approach to software development is the sustained engagement with end users. This helps avoid the delivery of a technical solution that is poorly matched to the skill set, interests, and needs of these "customers." Unanticipated problems can be uncovered early on, and dynamic allocation of priorities can help address them. Human factors play a large role in this approach, with an emphasis on effective communication across the enterprise.

Another distinctive feature of the Agile philosophy is to accomplish as much high-priority work as possible within the time and budget available, and declare success when the time and/or money run out. This contrasts with a requirements-driven approach, which is considered to have fallen short if those predetermined goals are not met.

I'm a strong proponent of the spiral/agile approach for software development. People seem to fall into two categories when it comes to thinking about software. Some take a top-down approach, designing data structures and middleware first. The second group tends to think from the inside out, focusing first on the core algorithmic problems that need to be solved. At the risk of over-generalizing and offending some readers, computer scientist and engineers primarily populate the first group, and scientists like me (who obsess about their data reduction algorithms) the second. The spiral approach balances these two factions, and drives the system to maturity. Also, after a couple of iterations the real problems begin to emerge, as opposed to (or perhaps in addition to) the ones you thought you'd have.

How extensible is this philosophy? The Agile approach is a current buzzword (Naslund and Kale 2020) in business management circles, often as a metaphor for nimble adaptation to changing circumstances. Is this a relabeling of longstanding good business practices, such as attention to team communication, placing a high value on delivered solutions that evolve in response to customer needs? I don't know.

There are inherent limits to using Agile methods for scientific/technical projects that have a substantial hardware aspect. It's not helpful to imagine building a succession of full-scale ten-meter telescopes, throwing them away at each iteration. But hardware prototyping iterations done at the subsystem level can be thought of as rotations through the Agile development cycle, undertaken to attain a fuller understanding of problems and their solutions. The nature of hardware work usually precludes multiple iterations happening on the two-week cycle time that Agile proponents favor. Moreover, the evolution of a project from conceptual design to preliminary design to final design constitutes three development spirals, except that the delivered items in each spiral are design and management documentation rather than working hardware.

THE LIFE CYCLE OF A BIG PROJECT

According to some pundits, the stages of a large project comprise the following:

1. Enthusiasm
2. Disillusionment
3. Panic
4. Search for the guilty
5. Punishment of the innocent
6. Praise and honor for nonparticipants

We aspire to avoid this fate. This chapter describes the typical sequence of stages and accompanying reviews for a major project. The exact definitions and review stages are slightly different for the NSF, NASA, and DOE, but the overall flavor is the same. Before turning to the agency-specific vocabulary differences, we'll sketch out the life cycle of a major project.

3.1 DEVELOPMENTAL STAGES OF A MAJOR PROJECT

The life cycle of any project starts with a good idea. This is then refined and tested through design iterations by the project team, and external peer review. The project management structure comes into existence in parallel, both supporting and reflecting the increasing clarity of the endeavor. The construction/development stage follows, with contingency being burned through as unexpected challenges arise. After a period of commissioning, an operational phase follows where the system delivers on its science promise. Once the system reaches the end of its useful life, it is decommissioned and the cycle begins anew.

This life cycle applies to projects at all scales, and technical project management tools can be used to good effect. Demonstrating an awareness of and competence with project management tools can build reviewer's confidence in the proposers carrying out the work described in a proposal.

The various funding agencies have somewhat different nomenclature and review stages for major projects, but the typical large-scale project follows the trajectory outlined below.

The start-to-finish timeline can be agonizingly long ("Atacama Large Millimeter Array" 2022). Site testing for the Atacama Large Millimeter Array (ALMA) telescope began in 1995. The agreement to split funding between Europe, Canada, and the United States was signed in 2003, and construction started that same year. The first science observations started in 2011. The first science paper appeared in 2012, seventeen years after the site testing.

In contrast, the Apollo program took only nine years from concept to landing people on the moon and returning them safely to Earth . . . using slide rules for performing many critical computations. Sigh.

3.1.1   THE GOOD-IDEA STAGE, AND COMPETITIVE ANALYSIS

Any project, big or not-so-big, starts with a good idea. This is the time to be harshly self-critical, and assess the merits of your plan in the context of the current competitive landscape. Almost all of modern science is done in competition with other groups. If there is no competition, domestic or foreign, it is likely worth understanding why! Take stock not only of the current competitive situation, but factor in the progress that others will make during the development stages of your project. Once your proposed system comes online, will it be competitive? What unique aspects will it have, and how long will that remain the case?

Talk the prospective project over with trusted colleagues, and identify its strengths and weaknesses. If you're not the originator but rather you're approached with the prospect of joining a new project, consider carefully how to spend your most precious resource, namely your time. Over my career I've shifted what I was working on to contribute to something else that came along, and I don't regret having done so.

3.1.2   CONSENSUS-BUILDING, TEAM-BUILDING, AND ADVOCACY

Good ideas build momentum, but they often need help. A handful of people acting as evangelists for a project is often necessary to generate the community interest needed to make big things happen. For complicated and expensive projects, it's often the case that one or more people dedicate the majority of their professional lives to building a critical mass of positive opinion and support. Talks at conferences, white papers, and foundational technical demonstrations can all contribute to this momentum, but it sometimes can take years for this to materialize. This is also a good time to review the scientific competitive landscape and consider whether a recon-figuration or consolidation is appropriate.

The tone and habits set at this stage will heavily influence what follows.

Expanding the team to incorporate needed expertise and disciplinary knowledge is common at this stage, but it's important that the group be functional and be able to handle and effectively resolve technical disagree-ments. The scientific leadership structure for the effort should start to emerge at this stage. The participants need to sort out how decisions will be made, and what governance structure suits them best. It's not too soon to talk explicitly about authorship and data access policies, to avoid collisions of unarticulated but incompatible assumptions.

The formation of a project consortium extends beyond gathering together individuals and research groups, to entire institutions. The initial engineering work needed to establish a strong foundation for a proposal can benefit from the staffing and expertise that are typically accessible to an institution rather than at a research group scale. Even at this early stage, it's not uncommon for institutions to formalize their working relationship by setting up a memorandum of understanding (MOU) that spells out their respective roles and early investments.

As the project team is assembled, explicitly seek out and invite a broadly representative cross section of the community and stakeholders.

3.1.3 DECADAL SURVEYS

An increasing number of scientific communities undertake a periodic "Decadal Survey" process that is a forum for building a consensus around a rank-ordered list of scientific priorities and goals. This often includes a ranked assessment of proposed major facilities and projects. Alignment with these community priorities increases the likelihood of a project being selected for funding. Proponents of major projects usually provide a combination of documents and presentations to these deliberations. It is increasingly the case that these decadal surveys are expected to make an assessment of the cost-effectiveness of proposed projects. Federal agen-cies are guided by, but are in no way bound to, these community prioriti-zation processes.

3.1.4 SUSTAINING COMMUNITY SUPPORT

The importance of clearly defining the intellectual scope of a major project and actively working to sustain community support is illustrated by recent tensions surrounding the Human Brain Project (HBP), a "Flagship Initiative" undertaken by the European Union. Originally described as a 1B Euro project to better understand the human brain, in 2014 a redefinition (or perhaps clarification?) of scope excluded major research communities of cognitive scientists. This led to a very public expression of disapproval in an open letter, and a subsequent rebuttal by HBP leadership.

This fragmentation is reminiscent of the pushback from physicists outside the subfield of particle physics, who voiced objections to the amount of money being spent on the Superconducting Super Collider. The very title of the book *Solid State Insurrection: How the Science of Substance Made American Physics Matter* by Joseph D. Martin (2018) conveys some of the undercurrents at the time. At the time of this writing, China is contemplating whether to embark on the construction of the next-generation particle collider, and similar discussions are under way there.

The need for sustained community support extends beyond just the scientific community. Large facilities are impactful in more than one way. Environmental concerns and cultural sensitivities require a thoughtful dialogue with local communities and other stakeholders. Astronomers, accustomed to being the "good guys," were caught off-guard when a groundswell of protests grew in opposition to the prospect of installing a major new telescope on the peak of Mauna Kea, on the big island of Hawaii. The issue attracted attention far and wide, posed a political challenge to the Governor, and went all the way to the Supreme Court of Hawaii for adjudication. This experience demonstrates that large scientific facilities can become contentious far from the walls of academia, and that a respectful and considerate resolution of concerns requires tact and diplomacy.

RESOURCE CONTENTION AND A BALANCED RESEARCH ECOSYSTEM

The funding that nations allocate for science is, in general, a scarce and oversubscribed resource. Apportioning these funds starts at a high-level distribution across fields and agencies, through national budgets. Within each agency there is a tension between allocating funds to large shared facilities and smaller-scale endeavors, right down to the sole-investigator research grant. Since not all scientists are equal users of shared facilities, or are not all equal partners in large collaborative projects, the big-science funding allocations can be a source of considerable frustration.

There are countless examples of this tension. Sometimes it's manifest in late-night gripe sessions among close colleagues, in other cases it boils over into full public view. The source of this frustration is easy to identify—the funding for a $500 million facility would cover 2,500 grants of $200,000. Which delivers the best return on investment? The answer depends on the value structure of who is asked.

It is vital that proponents of major projects understand, respect, and respond to this legitimate concern. This tension is particularly acute for the National Science Foundation (NSF). The NSF's Major Research Equipment and Facilities Construction (MREFC) process for major facilities does not currently provide a clear mechanism for funding the corresponding operating costs. The burden of the operating costs falls on the NSF division that is "lucky" enough to have secured an MREFC project, putting operations in tension with the PI grants program.

Proponents of large projects and shared facilities often argue, in order to help prevent the syndrome of a community circling the wagons and shooting inward, that science funding is not a zero-sum undertaking. Whether it is or not, the distribution of resources (or perhaps more importantly the perception of their allocation) both within and across stakeholder communities merits careful attention. *Solid State Insurrection* (Martin 2018) has a thoughtful discussion of the tensions surrounding big science and the funding it requires.

Even scientists within a facility-dependent subfield can disagree with the investment of portfolio vs. scale. Astronomers expressed concern about the extent to which cost and schedule overruns on the JWST squeezed out multiple flight opportunities for smaller, less expensive astrophysics missions. This quote from Alan Stern, ex–program officer at NASA, expresses the concern: "Are we going to turn off all the many existing astrophysics satellites and kill the support to analyze the data from them and stop building anything else, just so JWST can continue to overrun?" Stern said. "That's the question that the astrophysics community has to ask of itself, and that NASA should be asking."

The Particle Physics Prioritization Panel advises on priorities in high energy physics. Their 2014 "P5" report (Ritz et al. 2014) explicitly recommended, "Increase the fraction of the budget devoted to construction of new facilities." This careful phrasing, expressed as a fraction of the budget, explicitly prioritized investments in facilities at the expense of PI grants regardless of the total funding allocated to particle physics. This recommendation was presumably a result of the panel's consensus view that the field's future depends on access to cutting-edge facilities. Particle physics theorists objected. A number of theorists produced an open letter ("HET-HEPAP. Pdf" 2016) that read in part: "We formally request that a subpanel . . . be formed to investigate and better understand this damaging trend and to make recommendations to address its consequences and restore a thriving Theory program . . . and maintain the prominent and world-leading standing of US High Energy Theoretical Physics."

The lesson here is that advocates of a large project would do well to understand and address the legitimate concerns of the scientific community, and attempt to retain their goodwill during the (often protracted) project duration. Keeping to an assigned schedule and budget helps too.

3.1.5 PROPOSAL PREPARATION

Once the basic project goals have been clarified, it's not uncommon to submit one or more "R&D" proposals to secure the resources to perform basic design trade-off studies, and to achieve a level of engineering understanding sufficient to support a credible cost and schedule estimate. For some agencies this is a formal part of the big-project process. For example, NASA calls these Pre-Phase A and Phase A studies, while the NSF explicitly solicits "Design" proposals in the $600,000 to $6 million bracket. This R&D groundwork can take years for a billion-dollar scale project, especially if you factor in the time needed for proposal preparation and review.

This work culminates in a full project proposal. In some instances this might include funding for the scientific exploitation of the system, while in others the scope of the proposal might be for construction only. The NSF's MREFC process funds the construction but not the operations of facilities whose cost exceeds $70 million. The task of assembling and submitting a major collaborative project proposal is a serious managerial task in its own right, not to be underestimated.[1] Writing a compelling proposal narrative is not the hardest part. Assembling the requisite CVs, current-and-pending-support and conflict of interest sheets, the subcontract budgets and budget justifications, obtaining timely sign-offs from the sponsored project offices of the participating institutions, and wrangling the multitude of PDF files and associated documentation into a full proposal is a nontrivial and time-consuming task. If your team isn't able to accomplish this without undue stress, ask yourself how you'll manage to pull off the actual project, and reconsider.

A common practice is to assemble a "red team" of experts who have an arm's-length relationship to the project, to serve as an internal review team. They are asked to act as if they are a peer review panel, in the hope that their scrutiny will strengthen the proposal.

At many universities your colleagues in the grants and contracts office and those affiliated with the provost's office are usually excellent sources of advice, help, and support. Reaching out to them as well as agency program officers early on is likely to prove very useful.

If your vision for a big program entails substantial staffing increases in your facility or on your campus, or would impose other major demands on local infrastructure (computing, fabrication, etc.) then make sure the consequent space needs can be accommodated. It helps if you can point to whatever indirect cost revenue stream would be generated by the proposed activity, but bear in mind that indirect costs rarely (if ever) cover the full institutional costs of scientific research.

3.1.6   COMPETITIVE SELECTION AND REVIEW PRESENTATIONS

Federal support for science is highly oversubscribed and most proposals are not selected for funding. The down-selection process for large projects can include a combination of written submissions, scientific and technical presentations to review panels, and a site visit. Each of these is an opportunity to advertise the diversity, skills, and expertise of the project team. In my experience, the projects that are selected to move forward have a strong combination of (1) a compelling science case, that is well-aligned with agency and community priorities, (2) a credible technical execution plan, and (3) a cohesive team of key participants who earn the confidence of the peer review panel.

3.1.7   CONCEPTUAL DESIGN

The technical goal of this phase is to arrive at an overall conceptual design for the system, after an analysis of alternatives. The managerial aspects of the project should produce the high-level planning documents that will guide project execution. This includes an initial high-level Work

Breakdown Structure (see chapter 4) that identifies major tasks and their interdependencies and assigns them to elements of the project execution team. Top level technical risks are identified, with an initial assessment of likelihood, consequences, and priorities. The organizational structure of the project should be well-established at the end of the conceptual design phase. The safety plan should exist, at least in draft form.

The collaborative framework for the project team should also be well-established at the end of the conceptual design. The structure of the scientific collaboration should also be taking shape. An initial version of the key performance parameters should be produced.

This stage typically culminates in a Conceptual Design Review[2] (CoDR), where an external review panel assesses whether the project has made adequate progress to proceed to the next phase. As with subsequent reviews described below, the review panel will typically provide two kinds of recommendations. One is a mandatory list of corrective actions, which will be tracked through the execution of the project. The other is a list of suggestions for consideration by the project. An example of a mandatory corrective action might be "Ensure the organizational chart is fully populated with names for each key position, within the next sixty days." A suggestion for consideration by the project might be "consider the use of a CMOS sensor rather than a CCD for the fast-guiding camera on the second focal plane."

3.1.8 PRELIMINARY DESIGN

Once the overall system architecture has been settled, the next stage is arriving at an end-to-end preliminary design. The main transition that happens at this stage is establishing a baseline cost and schedule for the project, and finalizing the key performance parameters and the specifications and requirements for the various subsystems.

Reaching the preliminary design stage typically entails having completed 30–40 percent of the detailed design work, with a fully populated risk registry, and showing that the key personnel are in place and are functioning well as a team. The design work should have progressed to the point of having confidence in knowing what it will take to finish the project. There should be no unresolved substantial conceptual issues at the end of the preliminary design stage. The review panel is asked to assess the following:

- Are the performance parameters well-established, validated, and appropriately flowed down to the subsystem level?
- Will the proposed design accomplish the task?
- Are the project management structures in place?
- Is the team operating well, with clear lines of authority and communication?
- Is the budget well-justified, with appropriate contingency?

3.1.9 PRELIMINARY DESIGN REVIEW (PDR)

The Preliminary Design Review is an assessment of whether the project has achieved this state. Not all open technical issues have to be resolved, as long as there is a clear plan for picking between design alternatives, and the technical criteria and schedule for doing so is well-defined. Hand-waving answers to reviewer questions about outstanding issues are not likely to lead to a good review outcome.

The PDR is arguably the most impactful and in-depth review in the project cycle. It establishes the funding that will be allocated, and changing that number afterwards is (appropriately) very difficult. The PDR is typically a multiday affair, starting with an overview of the science case and an introduction to the project's conceptual design and high-level performance requirements. This is often followed by description of the project management, execution, oversight, and safety plans. This set of talks is often delivered by some combination of the Project Director, the Project Manager, and the Project Scientist. The review then often breaks into parallel sessions, by subsystem, with review panel members attending sessions that align with their areas of expertise. A shrewd review panel chair will have assigned review and writing tasks to individual review panel members at the kickoff executive session. The end of Day 1 brings the review panel together for an executive session, that often leads to a list of questions that are submitted to the project team. Some of these are questions of clarification, others are more pointed. These are often handed over to the project team at around 5:00 p.m., but because key project team members are expected to attend a quasi-social dinner with the review panel, work on answering them starts rather late on Night 1.

These parallel sessions often continue into Day 2, after the project team presents their responses to the questions that were posed at the end of Day 1. At some time in the afternoon of Day 2 the parallel sessions end, and the panel retires to an executive session, where they generate an "outbrief" presentation that summarizes their response to the charge they were given. The full project team is normally expected to attend the outbrief session. See chapter 5 for a more in-depth discussion of project reviews.

While the questions and comments from the reviewers are an important aspect of the PDR, the main value to the project comes from the review preparation and rehearsals, driven by a clear deadline.

3.1.10   FINAL DESIGN AND CONSTRUCTION

A substantial engineering effort occurs between the Preliminary Design Review and arriving at a complete system design. Once the project has decided between implementation alternatives and has completed a detailed design, a Final Design Review (FDR) determines whether the project is ready to proceed to construction. The FDR includes the full scope of the technical and managerial aspects of the effort, with a particular focus on whether the key performance parameters can be attained with the allocated resources. The FDR will also assess whether must-do recommendations from preceding reviews have been addressed.

If the FDR outcome is favorable, the project moves to the construction stage. If things are going well, the role of participating scientists is somewhat diminished at this point, and the scientists working on the project might want to look ahead and prepare for commissioning and verification. Periodic status reviews occur during this phase to assess technical status, cost, and schedule.

3.1.11   SYSTEM INTEGRATION AND COMMISSIONING

While it might be convenient to imagine a crisp transition from the construction phase to full science operations, it's usually more of a birthing process that takes some time. This transition constitutes the commissioning phase. The formal elements of this include:

- Full system integration, where the various subsystems come together
- Verification that the system meets its declared performance parameters
- Early science operations, collecting representative data and (ideally) pushing it through to the analysis stag.

Knitting together a complex system of systems invariably hits a few snags. Unanticipated interactions between the subsystems emerge.

Understanding, diagnosing, and resolving these issues requires drawing upon the collective experience and skills of the entire team. It is non-trivial for the team to shift from a subsystem-oriented view to a full-system perspective.

The formal deliverable for this commissioning stage is the verification of system performance. This tends to be centered on verifiable engineering parameters rather than, say, requiring that N science papers be produced.

3.1.12 SCIENCE VERIFICATION

Science verification is often the final step during commissioning before declaring the system operational. This goes one step beyond the demonstration of meeting engineering specifications; the science verification process drives data all the way through analysis methods to producing representative science results. There is a political delicacy here that the project needs to negotiate carefully. If the facility is built to serve community needs and aspirations, the hard-working scientists who have built the system are the obvious group to push through to early science results. But they run the risk of being criticized for "skimming the cream" without full community access to early data from the facility. Projects where the science users are fully integrated with the construction project don't face this challenge, but other configurations should be attentive to this issue.

A related issue is precisely when to declare the system ready for full operation, and community access to the data that flow from it. Providing community access too soon runs the risk of criticism due to residual instrumental artifacts, and other unresolved deficiencies. On the other hand, waiting until the system is fully mature, with an extended period of science verification, runs the risk of being criticized for limiting community access. Projects for which this is a potential concern would do well to think ahead about how to elegantly handle this transition.

The commissioning stage typically ends with an Operational Readiness Review (ORR), that assesses whether the system is sufficiently mature and well-documented to shift into "Operations." While a scientist might think of this as a continuous and ongoing process, from the federal agency's perspective the construction and the operation of the system use different kinds of money, allocated through different channels. It's not uncommon for the resolution of some lingering technical issues to be pushed into the operations phase.

3.1.13   OPERATIONAL PHASE

At long last, we reach the stage where the system enters into routine operations. There is a formal conclusion of the construction activity. The books are closed, victory is declared, and the construction team disbands. Many of the individuals carry their knowledge and experience forward, and join the operations team.

The details of what this entails depends on the character of the project, experiment, or facility. The organizational structure normally includes some manner of director, a tactically minded executive officer, appropriate engineering support staff, as well as administrative elements. A program advisory committee is a common oversight structure that will provide advice and plays an advocacy role on behalf of the user community. There are periodic performance reviews and oversight conducted by the cognizant federal agency or agencies.

Big facilities can be operated by a single university (like Princeton operating the Princeton Plasma Physics Lab), by consortia of universities (like the Keck Observatories), or by independent nonprofits such as the University Research Association (which operates Fermilab).

3.1.14   UPGRADES AND EXTENSIONS

It's not uncommon for major facilities to receive upgrades to extend their ability to deliver cutting edge science or to extend their reach into new applications. Examples of the latter include converting electron accelerators that were built for particle physics colliders into synchrotron light sources, using proton decay detectors as solar neutrino detectors, and using sensor systems that were built for national security for environmental monitoring.

Examples of facility life extension programs include the high-luminosity upgrade at the Large Hadron Collider, the Laser Interferometer Gravitational Observatory (LIGO) upgrades, and equipping both optical and radio telescopes with new instruments, electronics, and/or sensors. The extensive gene sequencing and analysis capability that was established for the Human Genome Project was subsequently brought to bear on a range of comparative genomics projects.

The timescale for the planning, design, and implementation of these upgrades can be long, so it's not uncommon to begin planning for upgrades even before a facility comes into full operation. This is a project phase where an innovative scientist can have a substantial impact. Leveraging the existing investment in a facility or project with a new operating mode, or by installing an upgraded instrument or detector, can be a very cost-effective way to extend the reach of a system.

NASA has a formalized process for evaluating whether to extend the funding for the operational phase of space missions.

3.1.15 DECOMMISSIONING

The annual operating cost (often called Operations and Maintenance, or O&M) of a major facility typically ranges between 4 percent and 10 percent of the original construction cost, and that fraction depends on a number of factors. A rather trivial arithmetic exercise will convince the reader that each decade of operations can require an investment comparable to the initial construction cost. At what point is science better served by taking that investment and directing it into a new facility or into entirely different science programs? Depending on the rate at which capability per dollar evolves for key technologies, at some point it's more cost-effective to terminate the operation of a facility, and replace it with a more capable next-generation system, even at a constant level of annual investment.

The scientific community tends to cling to existing facilities. An example of promoting a community discussion to grapple with this problem has been the succession of review panels appointed by the NSF's Division of Astronomical Sciences. In 2006 and 2012, the NSF assembled a Senior Review Panel and a Portfolio Review Panel, respectively (see "US NSF—MPS—AST—Division Portfolio Review" 2012), to provide advice on balancing the cost of operations of existing facilities against investing in new ones. This led to difficult conversations among stakeholders, with tensions that remain unresolved.

An informative example is the Arecibo radio dish in Puerto Rico. The 2006 Senior Review Panel commissioned by the NSF's Astronomy Division recommended termination of the NSF investment in operations at the end of that decade. This proved difficult to accomplish. Jobs are at stake, and in general the federal dollars that flow into congressional districts attract support on Capitol Hill. One impediment to the closure of extensive science facilities is site remediation. There are instances where the permitting process for land use requires that the site be returned to its prior state. This is a nontrivial undertaking, and it's sometimes cheaper to keep operating the facility than incur the costs associated with site remediation. Finally, although one might argue that overall science might benefit from closing old facilities to invest in new ones, there are user communities that lose out. Closing a radio telescope to invest in, say, a large-scale quantum computer is not likely to gain the support of radio astronomers.

In 2017, a decade after the Senior Review Panel recommendations, NSF scaled back their operations funding for Arecibo, shifting their annual operations support for radio astronomy to newer facilities such as the ALMA in Chile. The funding gap was partially filled by shifting/expanding Arecibo's mission to monitoring of near-Earth asteroids. This shifted the science focus to a different user community, with commensurate funding provided by NASA. In 2020 the Arecibo dish suffered considerable damage (Zastrow 2020) from broken cables. The combined effects of hurricane and earthquake damage led to the decision to terminate operations at Arecibo, fifty-seven years after it was commissioned.

NASA has terminated a variety of space missions, sometimes by driving the spacecraft into the surface of Mars (MESSENGER) or the moon (GRAIL). That's certainly one way to ensure the termination of the data collection phase of a project.

3.2   U.S. SCIENCE AGENCY PROJECT STAGES, REVIEWS, AND KEY DECISION POINTS

The various U.S. federal agencies that channel taxpayer dollars into science have different but somewhat overlapping mission areas, and distinct traditions and cultures. They each operate long-standing research centers or institutes (e.g., NASA Centers, DOE National Laboratories, NSF Centers, NSF Laboratories, NSF Observatories, NIH Centers, and NIH Institutes). Those establishments have more experience and expertise in project management than one typically finds in the university sector. Many consider this diversity of mission, focus and culture across the science funding agencies to be major asset of the U.S. science enterprise, as opposed to having a single "Department of Science." But one consequence of this heterogeneity is a rather complex set of somewhat inconsistent project stage definitions and project management vocabulary. Especially for projects that are jointly funded by more than one U.S. agency, it can be a challenge to disentangle and coordinate these different project stages, reviews and expectations.

This section summarizes the decision points and nomenclature used by the U.S. government nondefense research establishment.

I can't stress strongly enough how important it is to directly contact program officers early on if you're considering proposing a large project, be it facilities or a major collaborative science undertaking. Policies and programs are subject to change, and the program officers can provide excellent advice.

3.2.1   U.S. NATIONAL SCIENCE FOUNDATION (NSF) MAJOR PROJECT STAGES

The NSF categorizes instrumentation and facilities projects by their total cost. At the low end is the Major Research Instrumentation (MRI) category. At the high end is its Major Research Equipment and Facilities Construction (MRECF) program. In between are two Midscale proposal opportunities. Table 3.1 lists the breakdown of these facilities' funding opportunities at the NSF by cost category. Research proposals for large-scale projects (as opposed to requests for instrumentation or facilities construction) can be submitted as collaborative or consortium proposals. The distinction is described in section 6.4 below.

The Midscale R1 program has two components. Proposers can either request funding to carry out an entire project within the cost bracket, or they can request funding to carry out the preliminary design work needed to craft a mature proposal for the higher level R2 or MREFC funding.

The MREFC category of projects includes the most expensive facilities undertaken by the NSF. One challenge faced by the science community is that MREFC funding includes only the capital equipment costs. Operating

Table 3.1
NSF project categories

NSF program	Cost range
Major Research Instrumentation (MRI)	$100,000 to $4 million
Midscale R1	$6 million to $20 million
Midscale R2	$20 million to $70 million
MREFC	>$70 million, requires National Science Board approval

Note the gap in dollars between MRI and Midscale R1. It would seem that one should avoid an NSF-funded undertaking with a total project cost of $5 million!

Table 3.2

Major project stages as defined by the National Science Foundation (NSF)

NSF project stage	Associated review
Conceptual Design	Conceptual Design Review (CDR)
Preliminary Design	Preliminary Design Review (PDR)
Final Design	Final Design Review (FDR) National Science Board Approval if MREFC
Construction	Operations Review
Operations	

Table 3.3

U.S. Department of Energy project stages and reviews

DOE project stage	Associated review
Preconceptual Design and Mission Need	Independent Project Review, CD-0
Conceptual Design	Conceptual Design Review, CD-1
Preliminary Design	Preliminary Design Review, CD-2
Final Design	Final Design Review, CD-3
Construction	Operational Readiness Review, CD-4
Operations	

Also included are the Critical Decisions (CDs 0–4) that are often referred to in DOE projects.

expenses are shouldered by the appropriate science division(s) at the foundation, which incurs a tension with PI-scale research grants. Having a project accepted into the MREFC "queue" is a substantial undertaking, with a lead time of years to receive National Science Board approval and for funding to be requested and approved through the federal budget cycle.

Table 3.2 shows the formal major project stages as defined by the NSF, along with the associated reviews and decision points.

3.2.2 U.S. DEPARTMENT OF ENERGY MAJOR PROJECT STAGES

Table 3.3 shows the formal project stages for the U.S. DOE.

3.2.3 U.S. NATIONAL AERONAUTICS AND SPACE AGENCY (NASA) MAJOR PROJECT STAGES

NASA's project stages are laid out in table 3.4. One difference from the NSF and DOE is that NASA often funds multiple NASA Phase A studies, with

Table 3.4
NASA project stages and reviews

	NASA project stage	Associated review
Pre-Phase A	Concept Studies	Mission Concept Review
Phase A	Concept and Technology Development	Mission Definition Review
Phase B	Preliminary Design and Tech Completion	Preliminary Design Review
Phase C	Final Design and Fabrication	Critical Design Review System Integration Review
Phase D	Assembly, Integration, and Launch	Launch Readiness Review Operational Readiness Review
Phase E	Operations and Sustainment	Critical Event Readiness Review
Phase F	Closeout	

a down-selection that determines which will proceed for full support and eventual launch.

3.2.4 THE U.S. NATIONAL INSTITUTES OF HEALTH

The NIH comprises two dozen independent institutes and centers, with a research focus on disease and human health. The majority of these NIH units administer independent "extramural" (meaning funding provided to research conducted outside the NIH center or institute) grants programs. The NIH has less of a tradition, compared to NASA, NSF, and DOE, of constructing and operating major user facilities analogous to synchrotron light sources and telescopes.

3.3 EUROPEAN UNION MANAGEMENT OF MAJOR PROJECTS

There are equivalent processes in Europe for the selection and management of major science projects, and lessons to learn from experiences there. In particular, the EU criteria that are explicitly listed for project selection illustrate the nonscientific factors that bear upon those choices.

3.3.1 EUROPEAN SPACE AGENCY (ESA) PROJECT STAGES

Table 3.5 shows the project stages used by the European Space Agency. As with NASA, there is a funneling of candidate projects through a

Table 3.5

Project stages and reviews for the European Space Agency

Phase	ESA project stage	Associated review
0	Mission Analysis/Needs Identification	Mission Definition Review
A	Feasibility	Preliminary Requirements Review
B	Preliminary Definition	Preliminary Design Review
C	Detailed Definition	Critical Design Review
D	Qualification and Production	Acceptance Review and Operational Readiness Review
E	Utilization	End-of-Life Review
F	Disposal	Mission Close-out Review

succession of reviews and down-selection choices. The agency is guided by their Cosmic Visions 2015–2025 planning document ("ESA Science & Technology—Cosmic Vision" n.d.), which categorizes mission concepts into four broad categories (costs in 2007 currency, excluding launch):

- (S) Small missions, with total costs not to exceed 50M Euros
- (M) Medium class missions, with costs capped at 300M Euros
- (L) Large class mission, with costs capped at 650M Euros
- (F) Fast missions, executed on a rapid development timeline

ESA raised eyebrows in the science community in October 2020 with the cancellation of the SPICA mission, evidently due to independent cost estimates being inconsistent with the budgets provided by the SPICA project team. The agency has canceled or descoped missions in the past, but the science community has raised objections to the process that led to the SPICA decision.

3.3.2   HORIZON EUROPE

In 2019, the European Union's process for selecting its most expansive shared science projects was revised (Kupferschmidt 2019). The new process is termed Horizon Europe (Wallace 2020). Overall priorities are assessed through community input and discussion, leading to a strategic plan that will guide major European investments in projects and facilities ("Horizon Europe Strategic Plan" n.d.).

According to the European Commission, this plan will contain the following elements:

a. Key strategic orientations for R&I[3] support, including a description of expected impacts, cross-cluster issues and intervention areas covered;

b. Identification of co-funded and co-programmed European Partnerships—Institutionalised partnerships, based on Articles 185 and 187 TFEU, as well as the Knowledge and Innovation Communities of the EIT,[4] will follow the arrangements defined in their legislation;

c. Identification of Missions;

d. Areas for international cooperation, actions to be aligned with Research & Innovation of other nations and regions of the world at major scale, or actions to be carried out in cooperation with organisations in third countries;

e. Specific issues, such as the balance between research and innovation; the integration of Social Sciences and Humanities; the role of Key Enabling Technologies and strategic value chains; gender equality, including the integration of gender dimension in the R&I content; adherence to the highest ethics and integrity standards; priorities for dissemination and exploitation. ("Horizon Europe Strategic Plan" n.d.)

3.4   SELECTION CRITERIA FOR MAJOR PROJECTS

How do nations choose which large science projects merit the substantial investments needed to carry them out? Once the price tag for a project rises to the point of being a line item in the national budget, its future becomes subject to a wide range of factors, including:

• Scientific merit, as evidenced by peer review and prioritization
• Alignment with national technology development priorities
• Economic development plans (both national and regional)
• Issues of national prestige
• Workforce training and education objectives
• Diplomatic considerations, and desired international linkages
• Funding agency priorities and goals
• Resource availability, both financial and human capital
• National and regional political considerations
• The strength and persuasiveness of the project's leadership

While many of these considerations are outside the direct control of the scientific community, it's important to appreciate the complex nature of these decisions. The extent to which these major projects are seen as elements of both domestic and foreign policy can be seen from the factors enumerated by the EU's Horizon Europe program:

The Strategic Plan shall take into account an analysis covering at least the following elements:

a. Political, socio-economic and environmental drivers which are relevant for the EU and Member States' policy priorities;

b. The contribution of research and innovation to the realisation of EU policy objectives, while capitalizing on studies, other scientific evidence and relevant initiatives at EU and national level, including institutionalised partnerships;

c. Evidence-base resulting from foresight activities, S&T and innovation indicators, international developments such as the implementation of the SDGs[5] and feedback from implementation, including monitoring the implementation of specific measures with regard to widening participation and sharing excellence and participation of SMEs;[6]

d. Priorities with the potential to be implemented in synergy with other EU programmes;

e. A description of the various approaches for stakeholder consultation and citizen engagement as part of the work to develop Work Programmes;

f. Complementarity and synergies with planning of the KICs[7] of the EIT. (Horizon Europe Strategic Plan" n.d.)

The blend of scientific, economic, social, and political factors described above are indicative of the many considerations that bear upon the EU's selection of large projects at this scale.

AN OVERVIEW OF THE PROJECT MANAGEMENT TOOLKIT

This chapter will provide an introduction to the perspective, vocabulary, and methods used for developing and executing the task list, budget, and schedule—the technical project management toolkit. Understanding the challenges faced by professional project managers, and being familiar with the terms they use, will make you a better and more effective partner. Just as learning a few simple words and phrases can ease travel in a foreign land, mastering the bare essentials of project-speak will greatly enhance communication and mutual understanding.

The three interconnected core elements of project management are the task list, the budget, and the schedule. These each typically start with a rough plan on pieces of paper or a whiteboard, and evolve into a detailed documented project plan. Successful project execution requires active and agile management of the tasks, budget, and schedule. These core elements are supported by additional components including but not limited to the listing of project risks, the safety plan, and various memoranda of understanding.

4.1   ARE WE ON TRACK? SCHEDULING, MILESTONES, AND GANTT CHARTS

For an undertaking that is so complex that no single person can have full detailed knowledge of every element, we rely on breaking the endeavor into subtasks, with well-defined interfaces between them. These tasks are then assigned to subteams to carry out. The first step is to identify major elements and their subordinate items, and then figure out what a rational sequence of execution looks like. The generation of a structured hierarchical listing of all project work results in a Work Breakdown Structure, which gives a numerical designation and description to each work item. Arranging these into a sequence that takes prerequisites, dependencies, and time requirements into account produces a project schedule, which is often presented as a "Gantt chart" timeline.

For any project there is always a pacing item that determines the overall timeline for completion, and this can shift from one item to another over the course of project execution. The timeline that flows through the completion of these pacing items is called the "Critical Path" for the project. If we're building a toaster from scratch, the availability of bread is not likely to be on the critical path, but obtaining the heater element might be. Once the heating element is procured, completion of the fabrication of the housing might become the critical path item. Keeping track of what's on the critical path is a major focus area for technical managers. If things on the critical path fall behind, the entire project schedule slips to the right. Beware, however, the tendency to lose a sense of urgency for noncritical path tasks. If the pace of progress falls off too much, you can suddenly find them becoming the critical path item.

4.1.1   WORK BREAKDOWN STRUCTURE (WBS) AND TASK SEQUENCING

The project's Work Breakdown Structure is a hierarchical list of specific work packages, starting with high-level program elements that can have subtasks. Some of these WBS elements might include specific deliverables, either hardware or software, while others might be more process-oriented, such as Project Management. If there are more than a dozen or so high-level tasks, it is worth looking at whether some of them might be aggregated since it's hard for a person to assimilate a high-level overview if there are too many.

Each individual task has a numerical designator, such as 2.4.6. The leading digit is the highest index, called "level 1." It is followed by "level 2" and "level 3" indices, and so on.

These numerical designations indicate a task's location in the WBS structure. For a major project, each of these main WBS elements might have a dedicated project manager and subsystem engineering and science teams. You might hear people refer to the "level 2 manager" of a given subsystem. That means they have responsibility for that element and all that lie below it in the WBS hierarchy.

The highest-level entries of the WBS should correspond to the major subsystems for the endeavor. Examples might include:

- Data Acquisition Software
- Data Analysis Software
- Detector System
- Calibration System
- Database and User Interface
- System Engineering
- Project Administration
- Education and Outreach

Each item needs to be designed, developed, refined, and delivered. Going down the hierarchical WBS structure is akin to navigating a file system and directory structure on a computer. It's usually sufficient to have four to five levels of WBS hierarchy, but more complex systems might benefit from more layers.

An illustration of the structure of a typical WBS is shown in figure 4.1. Each element is associated with specific deliverable tasks or items.

It's essential that the WBS breakdown be complete—all tasks associated with successful project completion should appear somewhere in the WBS structure. Anything that isn't listed in the WBS will be declared out of the project's scope. To clarify dependencies and the extent to which tasks can execute in parallel, don't put unrelated tasks as subelements of a common parent. NASA has produced an excellent and accessible WBS Handbook (Terrell 2018) with good guidelines that are applicable to any project.

It's equally important that the WBS be understandable. Experience suggests that 500 is a sensible upper limit to the total number of tasks in the WBS structure. A more specific example of WBS task enumeration and definitions for a major project is shown in table 4.1.

One interesting choice is how to best include the software elements of a project. Is it better to have a single primary task for software, that contains all software for all subsystems as descendants, or should a subsystem's software appear as a subtask within that subsystem's WBS? Since management lines of authority ought to follow the WBS structure, unifying all the software under one umbrella has its advantages. On the other hand, subsystem managers will want to be able to control the software tasks for a deliverable for which they bear responsibility. Listing separate and independent software subtasks far down in the WBS structure incurs two main risks. First, a subsystem manager responsible for, say, a primarily mechanical engineering

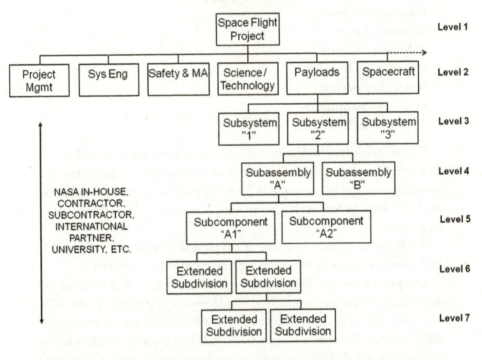

Figure 4.1
Work Breakdown Structure elements for a NASA flight project (Terrell 2018).

deliverable is likely to hire software developers with a mechanical engineering background, perspective, and favored toolkit. Second, fragmentation of the software effort can lead to complexity, for example if one team sticks to Python 2 while the rest of the project has adopted Python 3. Close coordination across the software elements of the project, with regular internal reviews, will help knit the disconnected efforts together.

Other elements of project management, such as cost accounting, risk management, and the assignment and alignment of roles, responsibilities, authority, and accountability should align with the WBS structure.

4.1.2 PRECEDENTS AND DEPENDENCIES

Once the main project elements and their hierarchy have been laid out, the next step is to assign dependencies. What tasks need to be completed before another can start? As an example, constituent subsystems need to be

Table 4.1
LCLS WBS DICTIONARY

WBS LEVEL No.					COST ACCOUNT	TITLE	DESCRIPTION
1	2	3	4	5			
1	06	06			10606	System Installation & Alignment	This element provides for System Installation and Alignment in all areas of the X-Ray Endstation System (Front End Enclosure, Near Hall, Tunnel, and Far Hall). Specifically, this includes controls, computer and network systems, safety systems, laser system, X-ray detector, and instrumentation and infrastructure for the atomic physics station. This also includes the integration of the X-Ray Endstation System with other components of the LCLS source, such as the LCLS timing and control system, vacuum system, and conventional facilities. Initial test activities are included in this section.
1	06	06	01		1060601	Front End Install	This element provides for System Installation and Alignment in the Front End Enclosure. Specifically, this includes controls, computer and network systems and safety systems.
1	06	06	02		1060602	Near Hall Install	This element provides for System Installation and Alignment in the Near Hall. Specifically, this includes controls, computer and network systems, safety systems, a complete instrument of a Atomic Physics Station with its experimental chambers and their vacuum components, and the laser system and optical transport. Initial test activities of these systems are included in this section.

Table 4.1 (continued)

WBS LEVEL No.					COST ACCOUNT	TITLE	DESCRIPTION
1	2	3	4	5			
1	06	06	02	01	106060201	Near Hall Install Controls	Closed.
1	06	06	02	02	106060202	Near Hall Install Network Fiber	This element covers the installation and cable termination of the network fiber for the FEE and the NEH. Also included is cable plant design, installation contract, and oversight.
1	06	06	02	04	106060204	NEH Install AMO Experiment	This element covers all Atomic Physics System installation in the Near Hall.
1	06	06	02	05	106060205	NEH Install Laser	This element covers the installation of the laser system and the optical transport in the Near Hall.
1	06	06	02	06	106060206	NEH Install PPS	This element covers all Personal Protection System installation, testing, and certifying in the Near Hall.
1	06	06	02	09	106060209	NEH Install Timing	This element covers the installation of the LBL timing system. It includes installation of the fiber trunk from sector 20 up to the NEH. Also included is cable plant design, installation contract, and oversight.
1	06	06	02	10	106060210	SXR Instrument Installation	This element includes labor and incidental materials for installation of the SXR instrument systems in the NEH.

A small portion of the WBS breakdown for the Linac Coherent Light Source (LCLS), constructed by the DOE at SLAC (from "LCLS WBS Dictionary"). This shows major subtask 1.6.6, "System Integration" and its descendants, down to five levels deep in the WBS structure. Note the cost account (CA) numbers in one-to-one correspondence with WBS tasks. *Source:* LCLS project 2009.

completed before they can be bolted together. This can be captured in the WBS by including in the attributes for each task the numerical identifiers of its prerequisites.

Identifying the start dates, durations, and necessary precedents for each task and its subelements will create an overall project flow. Figure 4.2 shows an example of this, for the admittedly contrived example of creating a piece of buttered toast with jam. There is a certain sequence of a subset of the WBS tasks that determines the pace of the overall project. Items along this trajectory are on the project's "critical path." Delay in any element of the critical path will delay the overall project.

In addition to tasks, the integrated project schedule can (and should) include intermediate milestones that are benchmarks for progress. These typically involve times when subsystems come together, or some other well-defined technical accomplishment such as completing the full initial system design.

4.1.3   BUILDING A SCHEDULE

The project representation in figure 4.2 is a "Gantt chart." Drawing our inspiration from the Wellington quote at the start of the book, we'll sketch out a notional project whose objective is to produce a piece of buttered toast with jam. Each task and subtask listed has a numerical designator, name, start time, estimated duration, end time, and list of predecessor tasks. The critical path (the sequence of tasks that determine the rate of progress to completion) is shown as the widest bars. For this particular example, the tasks associated with the procurement and disposal of the cow used to produce the butter are the primary factors that determine the time needed accomplish the task, given the proposed workflow.

There are some lessons embedded in figure 4.2. You might think you could just drive down to the local home appliance store and buy the toaster, refrigerator, knife, and plate so how could that aspect (Task 1) possibly take a week? Well . . . the institution's procurement system likely requires getting multiple quotes or bids from vendors, issuing purchase orders, documenting the process, and taking delivery. In parallel with this, the team can procure other elements such as the bread and jam, but only after finalizing the relevant requirements and specifications (strawberry jam? grape jelly? These things require heated discussions within the team . . .). Institutional viscosity and internal friction both tend to increase with scale-of-project.

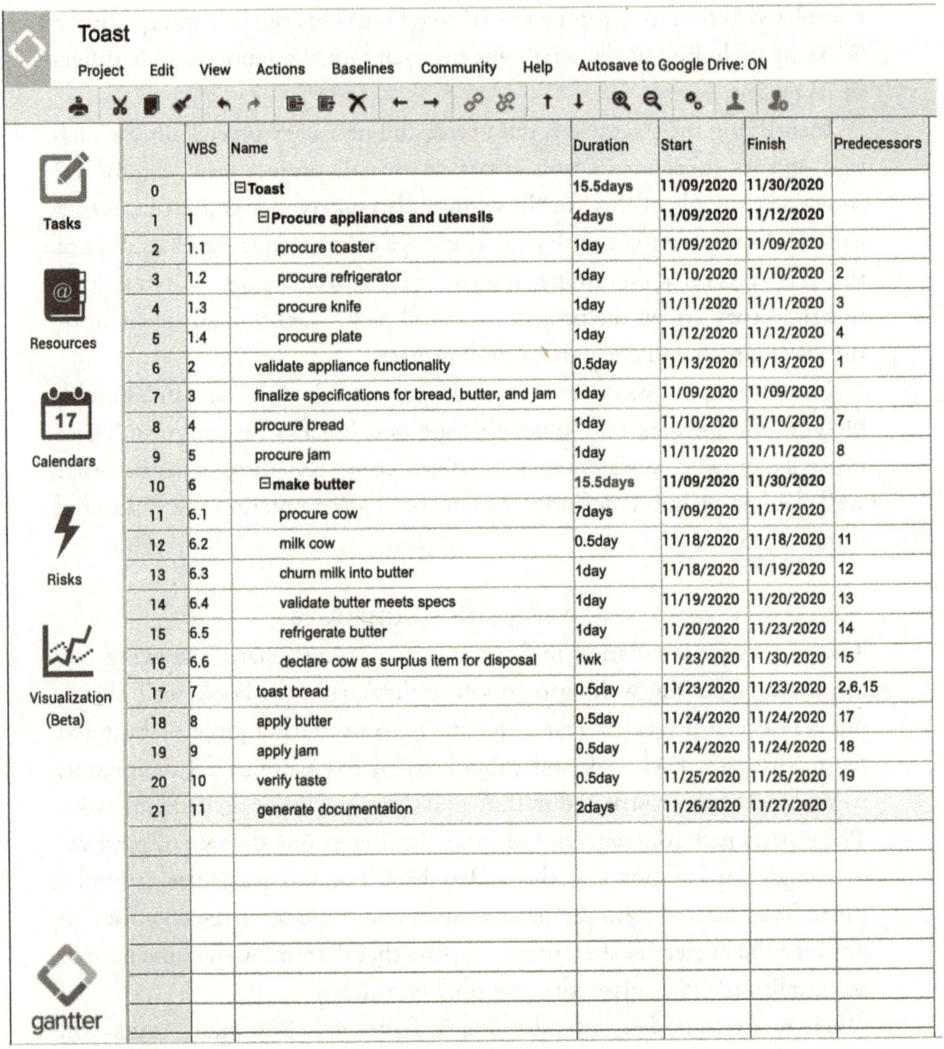

Toast

Project   Edit   View   Actions   Baselines   Community   Help   Autosave to Google Drive: ON

		WBS	Name	Duration	Start	Finish	Predecessors
Tasks	0		⊟Toast	15.5days	11/09/2020	11/30/2020	
	1	1	⊟Procure appliances and utensils	4days	11/09/2020	11/12/2020	
	2	1.1	procure toaster	1day	11/09/2020	11/09/2020	
	3	1.2	procure refrigerator	1day	11/10/2020	11/10/2020	2
	4	1.3	procure knife	1day	11/11/2020	11/11/2020	3
Resources	5	1.4	procure plate	1day	11/12/2020	11/12/2020	4
	6	2	validate appliance functionality	0.5day	11/13/2020	11/13/2020	1
	7	3	finalize specifications for bread, butter, and jam	1day	11/09/2020	11/09/2020	
	8	4	procure bread	1day	11/10/2020	11/10/2020	7
Calendars	9	5	procure jam	1day	11/11/2020	11/11/2020	8
	10	6	⊟make butter	15.5days	11/09/2020	11/30/2020	
	11	6.1	procure cow	7days	11/09/2020	11/17/2020	
	12	6.2	milk cow	0.5day	11/18/2020	11/18/2020	11
Risks	13	6.3	churn milk into butter	1day	11/18/2020	11/19/2020	12
	14	6.4	validate butter meets specs	1day	11/19/2020	11/20/2020	13
	15	6.5	refrigerate butter	1day	11/20/2020	11/23/2020	14
	16	6.6	declare cow as surplus item for disposal	1wk	11/23/2020	11/30/2020	15
Visualization (Beta)	17	7	toast bread	0.5day	11/23/2020	11/23/2020	2,6,15
	18	8	apply butter	0.5day	11/24/2020	11/24/2020	17
	19	9	apply jam	0.5day	11/24/2020	11/24/2020	18
	20	10	verify taste	0.5day	11/25/2020	11/25/2020	19
	21	11	generate documentation	2days	11/26/2020	11/27/2020	

gantter

Figure 4.2
An example of a Gantt chart for the making of buttered toast with jam. This project would have done better procuring butter rather than making it.

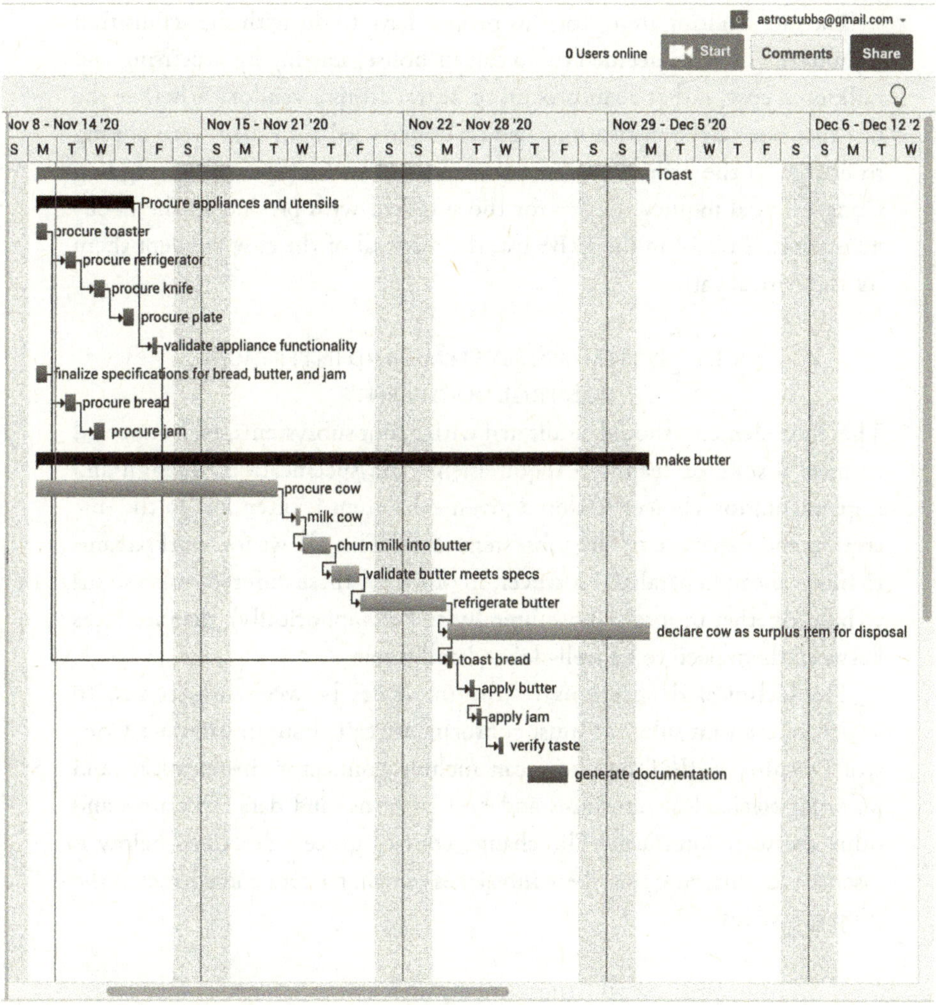
astrostubbs@gmail.com
0 Users online
Start
Comments
Share
Nov 8 - Nov 14 '20
Nov 15 - Nov 21 '20
Nov 22 - Nov 28 '20
Nov 29 - Dec 5 '20
Dec 6 - Dec 12 '2
S M T W T F S S M T W T F S S M T W T F S S M T W T F S S M T W
Toast
Procure appliances and utensils
procure toaster
procure refrigerator
procure knife
procure plate
validate appliance functionality
finalize specifications for bread, butter, and jam
procure bread
procure jam
make butter
procure cow
milk cow
churn milk into butter
validate butter meets specs
refrigerate butter
declare cow as surplus item for disposal
toast bread
apply butter
apply jam
verify taste
generate documentation

Figure 4.2 (continued)

The rate-limiting items for this project have to do with the acquisition of butter. The team decided to do this in-house, starting by procuring and milking a cow, rather than procuring butter from a vendor. Whether the fact that one of the institutions had an under-used barn they were hoping to charge to the project remains unclear, and is the topic of an ongoing Congressional inquiry. Lucky for those tasked with producing the documentation, Task 11 in the WBS list, the disposal of the cow has kept them off the critical path.

4.1.4   SYSTEMS, SUBSYSTEMS, AND INTERFACE CONTROL DOCUMENTS

The WBS elements should be aligned with major subsystems, each expected to meet a suite of technical requirements and specifications. Design and implementation choices within a given subsystem is often left to the discretion and expertise of the subsystem team. This allows for workstreams to move ahead in parallel. However, in order for these different subsystems to bolt together properly (both literally and metaphorically), the interfaces between them need to be well-defined and stable.

The technical descriptions of the interfaces between subsystems, to which all relevant subunits must conform, are called out in Interface Control Documents (ICDs). These can include connector specifications and pinouts, mechanical interfaces and bolt patterns, and data structures and other software interfaces. The change control process described below is essential for ensuring that these interfaces remain current and correct as the project evolves.

4.1.5   SCHEDULING SOFTWARE AND TOOLS

There is a spectrum of management tools that can be used to generate and curate the WBS, schedule, status, and resource allocation for a project. The ones named here are meant as illustrative examples of classes of project management tools. These project management tools can include the ability to assign resources (e.g., people, money, time) to specific tasks, with support for embedded data structures that enable Earned Value tracking (described below) and maintaining a risk registry. The trick is to pick the right scale for your project, balancing ease of use against complexity.

Enterprise-scale solutions such as Primavera P6 from Oracle can support simultaneous users, track thousands of tasks, and make links to external databases, with the ability to do scenario "what-if" planning. It's not the kind of thing people typically run on a laptop.

One desktop-scale project management tool is the Microsoft Project program. The locally run version is only available for the Windows operating system, but there is a cloud-based implementation.

The Gantt chart in figure 4.2 was made with a Google Chrome add-on called Gantter, which is a lightweight tool that readily allows for sharing online and collaborative access and editing. Numerous similar tools exist, with various combinations of ease of use, complexity, and features.

4.2 WHAT WILL IT COST? ARE WE GOING BROKE? BUDGETING AND COST TRACKING

Cost estimation for a big project is an iterative process, that entails feedback between design and pricing, with compromise as a key element. As the system evolves toward a more mature and well-understood design, and the dominant risks are retired or mitigated through analysis, verification, and prototyping, the 10–90 percent confidence cost envelope should shrink. The U.S. Government Accountability Office (GAO) publishes a "best-practices" guide (Young, Brown, and Blockwood 2020) that draws upon experience across the various agencies. This is a very useful reference for cost estimation. Other agencies have their own handbooks.

The definition of total project cost depends on what is included in the calculation. One might imagine this is a crisply defined quantity, but alas it isn't. How, for example, does one include in-kind contributions[1] from foreign partners?

Coming up with a credible, realistic estimate of how much it will cost to do something that's never been done before is difficult, due to the inherent uncertainties. Moreover, most of the incentives drive toward cost underestimates. The project team wants for their proposal to look attractive and affordable. As described earlier, people tend to underestimate the amount of time it really takes to get things done and fully documented, especially when this work is being done with administrative overheads that are imposed by the review and oversight processes that are normally imposed on major projects.

There are three important points that you need to absorb about project cost estimation:

1. Essentially all incentives drive in the direction of initially underestimating project costs: lower-cost projects are more likely to be funded in a highly competitive scientific ecosystem, we all tend to underestimate the time and difficulty of actually finishing things, and project proponents tend to underrate risks, both technical and to schedule.
2. The historical record of cost overruns on major projects is sobering, across essentially all sectors of the economy. Science projects are no exception.
3. Delays incur an increase in total project cost. Annual escalation of the cost of labor, materials, and services is one reason. But arguably more impactful is the desire to keep a project team intact, and the "marching army" salary cost adds to the project total if the schedule slips.

If you're at the early stage of a major project, I encourage you to learn some of the history (see, e.g., *Megaprojects and Risk: An Anatomy of Ambition* [Flyvbjerg, Bruzelius, and Rothengatter 2003] and *Tunnel Visions: The Rise and Fall of the Superconducting Supercollider* [Riordan, Hoddeson, and Kolb 2015]) of cost overruns on major projects and consider the human factors that have contributed to a sobering historical track record of overoptimism. Table 4.2 illustrates some examples of projects whose costs exceeded their initial estimates. As you make a good-faith effort to produce a cost estimate for a facility or for a large-team science project, take those historical lessons into account.

Table 4.2
Some sobering examples of initial cost underestimation

Project	Cost overrun compared to initial estimate	Status
Suez Canal (1860s)	2000%	Completed
Boston "Big Dig" highway project (1990s)	190%	Completed
James Webb Space Telescope (2000s–2020s)	1000%	Completed
Superconducting Super Collider (1980s–1990s)	Projected to be >200% when canceled	Canceled

Source: Suez Canal and Boston highway data from Flyvbjerg, Bruzelius, and Rothengatter 2003; SSC from Riordan, Hoddeson, and Kolb 2015; JWST from Billings 2010 and U.S. GAO 2021.

These are among the reasons why some agencies subject major proposals to an independent assessment of the validity of a project proposal's cost section. This is nothing new—the proposal that Stanford submitted to the government in 1958 for the construction of the linear accelerator that eventually became SLAC was subjected to independent cost review by two engineering firms (Kirk 1970).

4.2.1 ELEMENTS OF A PROJECT BUDGET

Budget elements for a major project comprise the same categories that appear in smaller-scale proposals, but with more people, more stuff, and a correspondingly larger bottom line. Line items typically include:

- personnel costs—salary and benefits for people working on the project
- capital equipment—long-lived items, typically that exceed a cost of a few thousand dollars
- supplies and materials—expendable items like connectors, wire, etc.
- travel costs—domestic and international travel needed to execute the project
- technical services—machine shop, computer time, and engineering work if not paid through salaried employees
- indirect costs—"overhead" costs to cover institutional expenses for laboratory and office space, and administrative services

As a project moves from conceptual to preliminary design to construction, a cost "baseline" is established, which should be interpreted as a "not-to-exceed" total project cost. The final stage of establishing this entails generating the equivalent of many three-ring binders filled with specific vendor quotes for items the project will need to procure, ranging from power supplies to computers to sensors to technical services. These "cost books" and the parts list for the project determine how much money will be spent on buying things.

If the work is going to be conducted at more than one institution, subcontracts are involved. This means that the sponsored projects and grants administration components of multiple institutions need to coordinate their budgets and signature processes, and this can take a considerable amount of time. Be sure to engage your local grants office in this process as early as possible and keep them abreast of developments.

Cost estimation for the hardware elements of a project is challenging. But once the work has been broken down into well-defined subtasks, one can draw upon analogous prior tasks. Precision machining, electrical engineering, and systems integration can be estimated from either a parametric approach or using a bottom-up cumulative estimate.

4.2.2   METHODS OF COST ESTIMATION

Fortunately, most projects that are under consideration have precedents that can be used as a starting point for cost estimation, and the people who do this kind of work tend to move from one project to the next. So, rather than starting with a blank sheet of paper, one can work forward from prior experience. For satellite systems there is enough accumulated knowledge and experience that parametric cost estimation tools are often used to check the project team's estimates against the heritage of prior satellite construction experience. Parameters like the mass of the system, the aperture of the optical system (if there is one), the satellite bus agility, and the downlink data rate are examples of factors that are used as inputs to parametric cost models. Similar cost scaling laws can be derived for ground-based optical telescopes. Any project that claims they have to break away from these scaling laws would have to show persuasive evidence to support that assertion.

The three main methods of generating a project cost estimate are (1) bottom-up estimation based on a detailed project task list, (2) argument-by-analogy, based on one or more similar undertakings and accounting for the differences, and (3) parametric cost models based on lines of code, pounds of machined metal, and various quantitative descriptors of technology content. To the extent that more than one method can be brought to bear, one can get a sense of the uncertainty in the cost project's estimate.

TASK-BASED COST ESTIMATION   The task-based cost estimation method entails assigning the labor, materials, and other costs to each element of the Work Breakdown Structure, and then summing up these costs across the entire task structure. This is likely to require the participation of multiple individuals, spanning the range of expertise across the proposed enterprise. Doing this in a consistent fashion will require a common approach, especially regarding contingency and "padding" of cost estimates. If at each level of aggregation, moving up through the WBS structure, the managers insert conservative budget padding, the estimate is not likely to be accurate.

One major advantage of performing this task-based cost estimation is the subsequent use of value assignments for Earned Value Management (EVM), as described below. Since EVM is the currently preferred method for tracking project execution, you'll eventually have to assign a dollar value to each task. So (subject to staffing constraints) it makes sense to get started on that early.

COSTING-BY-ANALOGY   This method uses one or more closely related prior examples as the basis for cost estimation, by computing only the cost *differential* due to differences between the two. The pitfalls here include not accounting for all the enhancements and upgrades, as well assuming that items which were procured in the past can be readily obtained in the future.

PARAMETRIC COST ESTIMATION   The parametric cost estimation process uses a multiparameter description of system attributes and a database of prior project cost history to construct an estimate on the basis of a multiparameter fitting process. A simple illustration of this idea is shown in figure 4.3. Prices (MSRP) and weights for a rather random assortment of passenger vehicles were obtained for seven vehicles.[2] This information implies that (at least for these vehicles) the price of a mass-produced vehicle can estimated based simply on its weight, using a second order fit. A more sophisticated version might include luxury class, country of production, and other factors.

If one were tasked with estimating the price for a new mass-produced vehicle, the parametric approach would have two stages. First, estimate the final mass of the vehicle, then use the parametric model to estimate the price.

This approach is frequently applied to satellite projects, where a substantial database of prior projects exists, with parameters that include vehicle weight, power, sophistication of instrumentation, downlink data rate, and other factors. For space vehicles there are spreadsheets that can be used to generate cost estimates using those parameters as inputs. The Aerospace Corporation has substantial experience in this domain.

ARRIVING AT AN INITIAL COST ESTIMATE   There is an entire cohort of project management professionals who specialize in cost estimation for

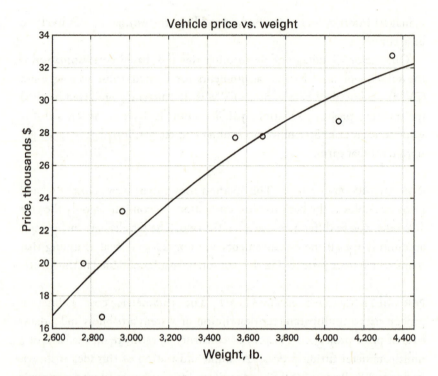

Figure 4.3
Retail price of passenger vehicles vs. weight. A quadratic fit provides a description of
price vs. weight that matches the data to a fractional precision of about 10 percent. If
one is tasked with estimating the price of a new mass-produced vehicle, a parametric
estimation like this is one possible approach. Including a "luxury" factor would likely
enhance the fidelity of this parametric predictor of vehicle cost.

technical projects. Companies that submit proposals for government contracts are highly motivated to get this right. If they submit a fixed-price bid
with a cost estimate that's too low and they're awarded the contract, they
can lose a lot of money. On the other hand, submitting a proposal with a
cost estimate that is far above other competing proposals means they won't
get the award.[3] The NASA centers and DOE national laboratories are institutions in the national R&D complex that have considerable experience
and expertise in cost estimation. Universities are less likely to have someone
on their permanent staff with broad experience in this domain.

If you're putting together an ambitious large-scale proposal and you
don't have an experienced professional overseeing this aspect, consider
either augmenting the team with an institution that has this as a core

competency, or else engage an external consultant to assist with this important task.

ESCALATION AND INFLATION Doing things in the future typically costs more than it would take to do them today. One contributing factor is inflation. Purchasing power erodes at the rate of inflation, which is computed on the basis of items most customers buy. The cumulative effect of inflation is significant. A "flat" budget over a decade with 3 percent annual inflation has about 75 percent of the purchasing power in the final year as it did in the first.

Escalation is a slightly different concept, which tracks how the cost of goods and services go up even if consumer-price-index inflation were zero. Here's a concrete example: The salary cost for a project usually goes up from year to year, simply because the people working on the project get well-deserved raises, even with no inflation happening in the local economy. On the other hand, the technical capacity and capability that a dollar will buy tends to increase over time. The cost of desktop computers has been about the same over the past ten years but my goodness they are far more capable.

Presentations and interpretation of fiscal information and projections must distinguish clearly how inflation and escalation are taken into account, and this has a vocabulary all its own. Apples-to-apples cost comparisons are done in "real dollars" or "constant dollars" or "inflation-adjusted dollars." Actual expenditures are presented in "then-year" or "current dollars." One clear way to present this information is to adopt a "base-year" for all budgeting, and express all cost figures in base-year dollars.

Reviewers and agencies expect escalation to be included in project budgeting. One challenge is to choose, at the early stages of a project, appropriate cost escalation factors. This is a good topic for discussion with your sponsored projects office and agency program officials.

4.2.3 BUDGETING FOR SOFTWARE

Estimating the labor and time needed to produce the software for a project is particularly challenging. A cost estimation manual from the Government Accounting Office (GAO) dedicates Appendix V to the challenges of estimating software costs (Young, Brown, and Blockwood 2020). Here are some excerpts from that document:

During the ten-year period between the first flights of the Boeing 757 and of the Boeing 777, software code increased 21-fold and the software development cost increased by a factor of more than 28-fold. On the military side, the first F-35 was flown 32 years after the first F-16 with 177 times as much computer code and the software development effort cost totaling almost 300 times as much.

In 2018, the Project Management Institute reported that 14 percent of IT projects were deemed failures and that on average 9 percent of every dollar spent on IT projects was wasted. Among the projects surveyed, 56 percent reported on-time completion while 60 percent were completed within budget.

There are countless examples of projects that have underestimated, or grossly underestimated, the effort needed to generate validated, verified, documented code that meets requirements. The consequences can be a "soft fail" where the software shortcomings mean the project did not reach its full potential, or a "hard fail" where the entire project collapses. Good communication is key, as illustrated in figure 4.4. Examples of the latter include some spectacular failures of software planning and execution. One textbook case was the Federal Aviation Administration's (FAA) program to upgrade the nation's air traffic control system to something called the Advanced Automation System (AAS). Here are some excerpts from the auditor report (U.S. DOT 1998) that followed the collapse of the project:

> Software development proved to be the Achilles' heel of the AAS program.
>
> Of the $2.6 billion spent on the total program, we estimate the total loss to the Federal Government to be about $1.5 billion, excluding FAA personnel costs.
>
> At each point when the program slipped, it became more clear that FAA and IBM drastically underestimated the complexity of developing AAS software.

The tools for managing large software efforts have advanced since the 1990s, when the FAA fiasco occurred. Spiral development (where rough-cut, end-to-end solutions are iterated toward a final solution) and "Agile" methods explicitly try to maximize the efficiency of building working code. There is a palpable tension between extensive top-down design, defining requirements, and writing documentation versus just getting something crude working quickly to fully understand the scope of the problem, and working outward from there.

WE NEED TO MAKE 500 HOLES IN THAT WALL, SO I'VE BUILT THIS AUTOMATIC DRILL. IT USES ELEGANT PRECISION GEARS TO CONTINUALLY ADJUST ITS TORQUE AND SPEED AS NEEDED.

GREAT, IT'S THE PERFECT WEIGHT! WE'LL LOAD 500 OF THEM INTO THE CANNON WE MADE AND SHOOT THEM AT THE WALL.

HOW SOFTWARE DEVELOPMENT WORKS

Figure 4.4
Software development coordination is a good thing. Graphic from "Software Development" (Munroe, n.d.).

Of all the elements that are needed to make a big project succeed, in my experience the software aspect seems to have the biggest variation in team productivity. I've seen instances where a highly capable, experienced, and motivated team of six people outperforms groups ten times their size.

For many projects, the software aspect essentially ends up as a level-of-effort activity. A certain number of people get hired onto the project, and they work until the allocated resources are expended, and the scope of the software deliverables is adjusted to fit the time and budget constraints. That suggests that it's a good idea to define a core of must-do elements (data acquisition, storage in a database, tracking calibration metadata and provenance . . .) and items that could be descoped without threatening the key project deliverables. If the software team manages to get it all done, great. But knowing in advance which things to scale back or defer, should it prove necessary, will speed the decision-making process.

Bottom line on software cost estimation: My advice is to make a very conservative and pessimistic estimate for the software development effort needed for a project, then triple it. Then add a hefty amount of contingency, as much as 30–40 percent of the software budget. Then go attract the very best people you can find to help get things done.

4.2.4   TRACKING AND CONTROLLING COSTS— EARNED VALUE MANAGEMENT

The financial administration of the project requires tracking the rate of expenditure of money against progress toward completion. This in turn requires that the technical team deliver ongoing updates about status and assess the remaining work. The basic idea behind "Earned Value Management" is to cast both cost and schedule information in units of dollars. This applies to both a snapshot of current project status as well as projections into the future. Project managers use this information to compute, on an ongoing basis, the Estimate to Complete (ETC) and Estimate at Completion (EAC). These are forecasts that take the current status and extrapolate how much money is needed to finish (ETC) and the total that will have been expended at completion (EAC).

Earned Value Management (EVM) is a methodology that tracks the "value" generated by effort and expenditures. In effect it defines a new unit for project productivity, "Present Value," and assesses how well the team is "earning" credit in this framework. Implementing EVM is somewhat cumbersome, since it requires assigning an appropriate dollar value to each element of the Work Breakdown Structure and regular tracking of partial completion status. The sum of the assigned Values across the WBS equals the budget allocated to the tasks in the WBS.

Figure 4.5 shows the basic idea. The vertical axis is in units of dollars. The Actual Cost (AC) profile shows the rate of expenditure of actual dollars. The Earned Value (EV) line shows how the project is accumulating value over time. The Planned Value (PV) line shows the desired project execution trajectory.

COST VARIANCE AND COST PERFORMANCE INDEX   Any discrepancy between Actual Cost and Earned Value is defined in units of dollars as

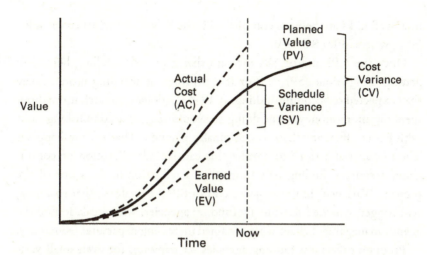

Figure 4.5
Earned Value Management. The basic idea is to cast both cost and schedule information onto a common "value" basis. The project accrues Earned Value as progress is made against WBS tasks, and money is spent at a rate captured by Actual Cost. Discrepancies between reality and the Planned Value are captured as "variances."

the Cost Variance (CV). The sign convention is that a negative Cost Variance is bad, and a positive cost variance is good, that is, $CV = EV - AC$, in units of dollars. It's somewhat more informative to cast this information as a dimensionless number, so a normalized Cost Performance Index (CPI) is defined as $CPI = CV / AC$. Program officers at funding agencies often track CPI as a function of time to assess progress on projects in their portfolio. Quoting from the U.S. Department of Energy's Earned Value Management System and Project Analysis Standard Operating Procedure document: "A cumulative index of less than 0.95 is used as an early warning indicator of cost increase and should be investigated" (U.S. DOE 2020).

SCHEDULE VARIANCE AND SCHEDULE PERFORMANCE INDEX   A related concept is Schedule Variance (SV). One might think this would be the horizontal distance, in units of time, between reality (EV) and the project plan (PV). But SV is defined as any gap, positive or negative, between the desired and actual earned value. So, $SV = EV - PV$ also has the sign convention where a negative SV means the project has fallen behind expectations for progress. The normalized dimensionless parameter used to track how well tasks are being accomplished is the Schedule Performance Index (SPI), where SPI = SV / PV.

Once the CPI and/or SPI indicate that a project is falling behind in getting things done (SPI < 1) or that a project is spending more money than expected to accomplish things (CPI < 1), it's hard to catch up without spending more money, since doing so would require accomplishing tasks with better efficiency than was originally planned. That seldom happens. The managerial actions to correct this can include allocation of contingency resources, finding new money, or a reduction in the scope of the project. "Variance" in the project's CPI or SPI (i.e., values other than one) will trigger increased scrutiny by funding agencies, with escalation up the agency management chain with sustained increasing departures from unity.

Program officers in funding agencies set tripwires for even small variances in CPI or SPI, for two reasons. First, there is an unavoidable latency between the instantaneous current status of a project and what is captured in the Earned Value information, so things are likely worse than the CPI and SPI numbers reflect. Second, identifying issues early allows for more leverage on corrective actions.

SCHEDULE SLIPS AND DELAYS   A somewhat more intuitive indicator of whether a project is on track is whether tasks are getting finished at the anticipated rate. Any lag between the planned and actual completion date is a schedule slip. Setting integration milestones as benchmarks or milestones in the project plan, at specific dates, is a good way to highlight for everyone in a totally accessible way whether the project is keeping up with its plan. Delays in execution eat into schedule contingency.

When deadlines are missed, it's common for participants to have two responses: (1) provide a technical rationale, with the impression that it somehow justifies the delay, and/or (2) criticize the plan and timeline. While it is sensible to adjust the execution plan as the project gains experience and the design matures, it's vital to realize that unless delays are made up later, the project runs the risk of going over budget.

TRACKING SOFTWARE PROGRESS Tracking the status of software development over the course of the project is as challenging as budgeting for it in the first place. Establishing clear milestones and hitting the project's fire alarm if they start to slip is the best suggestion I can offer.

4.2.5   TOOLS FOR BUDGETING AND COST MONITORING

The budget-construction phase usually doesn't require more than a spreadsheet, albeit sometimes a rather complicated one. This could be a shared-editing document in the cloud, or a file that is emailed around. As with all things, attention to version control and file naming conventions will avoid trouble.

The cost-tracking aspect of project execution requires making a choice about the appropriate level of complexity to invoke. Keeping track of a full-blown Earned Value Management System (EVMS) entails considerable project management overhead. An intermediate scale of cost oversight is to use clearly defined intermediate technical milestones and take stock of the funding that remains as those milestones are achieved. An alignment of both technical and fiscal authority, accountability, and responsibility at the various management levels across the project will help midlevel managers execute their portion on time and on budget.

A common approach is to assign a Cost Account Manager (CAM) to appropriate sized elements of the Work Breakdown Structure. This aligns the financial oversight with the overall project structure.

4.3   WHAT DID WE BUILD? HOW WELL DOES IT WORK? TECHNICAL DOCUMENTATION

One substantial organizational challenge is to generate, update, and curate the reams of documentation that accompany a large project. This includes:

- Cost justification materials, called "cost books"
- Proposal documentation, conflict of interest declarations, MOUs, and other agreements
- Project management documents such as project execution plan, safety plan, management plan, and verification plan
- Financial documents
- Personnel files
- Procurement documentation: quotes, bids, and invoices
- Technical requirements and specifications with flow-down to subsystems
- Interface control documents
- Mechanical designs and drawings
- Circuit diagrams and board layouts

- Parts lists
- WBS and Gantt charts
- Project risk registry
- Technical documentation and performance verification
- Operations and service manuals
- Software source code

Most of these are living documents, that evolve (in a controlled fashion) over the course of the project. That means a version control system needs to be established. The active curation of this heterogeneous library of information is nontrivial. Establishing a searchable up-to-date document repository that can productively be used by the project team is important. For large projects this can be a full-time job for someone.

Pitfalls include not establishing naming conventions and not defining an organizational structure for mechanical drawings, electronics, and software documentation. Scrambling to put things into some logical structure after the fact is painful. Give up on trying to get engineers to all conform to one suite of CAD software. People are wedded to the tools they like, and the documentation control system should accommodate different file types.

Repositories like Confluence, DocuShare, Google Drive, SharePoint, and the like can be used to upload and store documents in a hierarchical file system, but in my experience they all tend to be cumbersome when one tries to search for files or documents that pertain a particular topic. In addition to the design documents, it's good practice to generate and log technical notes, internal technical reports, and decision memoranda. PowerPoint is no substitute for hard-nosed technical documentation, so let's all try to resist the trend in that direction.

A distinct but related problem is archiving data—calibration and verification data, early science verification data, and their associated configuration and setup files. You should expect for the project management team to be emphatic in their insistence that the participating scientists conform to metadata standards and documentation discipline.

4.4   WE GOT THIS, WHAT COULD POSSIBLY GO WRONG? RISK MITIGATION

While it's nice to imagine that a complex project will unfold precisely as planned, this is seldom the case. An important element of technical management is to identify and attempt to quantify the various risk elements in that might pose problems as the project matures. A rank-ordered list of these various risks, taking into account their consequences on cost and schedule should they come to pass, allows for an orderly and rational allocation of resources. A "risk mitigation" strategy is a layered approach that uses simulations, prototypes, and analyses of alternatives to reduce the overall risks as the project proceeds.

Much like the Earned Value approach described above, it's common to ascribe a dollar value to both the direct cost and schedule impacts that would result from a risk turning into reality.

4.4.1   RISK REGISTRY

A pessimist is an optimist in possession of the facts.

The risk registry contains a listing of potential problems that might arise. An illustrative example is shown in table 4.3. Each possible adverse event is assigned a likelihood of occurrence, along with estimates of the impact it would have on both cost and schedule. The schedule impact is often converted into an equivalent dollar amount, in order to put both the cost and schedule consequence onto the same system of units.

Project managers use a risk registry to produce a rank-ordered list of concerns, usually taking the product of a quantitative likelihood estimate times the risk's consequence. "Risk mitigation" or "buying down risk" takes place early in the project cycle, where an investment of funding and effort is used to reduce the project's exposure to the most consequential concerns. Refined design and finite element simulations can often address

Table 4.3

An example of a risk registry

Risk	Likelihood (%)	Cost impact ($ million)	Schedule impact ($ million equivalent)	Risk factor
Vacuum failure	2	3.2	5.4	17.2
Software delay	10	0	4.4	44

The first column shows the potential event, the second column is the likelihood, the third column is the cost impact (hardware replacement), the third column is the dollar-equivalent value of the schedule loss, and the final column is an overall relative risk rank factor, computed as RF = likelihood * (cost impact + schedule impact).

mechanical engineering concerns. Prototyping of software and hardware can increase confidence in system components. This leads to adding columns to the risk registry database, showing the post-mitigation residual risk. The project management team should strive to bring the postmitigation risks down to an acceptable level. The project plan should include sufficient "contingency" resources to achieve this.

This approach to risk assessment has two advantages. First, it gives us a way to generate a rank-ordered list of concerns, on a semi-quantitative basis that derives from the expert judgment of people who know the project well. Second, it provides a way to assess whether sufficient contingency (in both dollars and schedule) exists, compared to the totals in the final column in table 4.3.

A common mistake is to underestimate the overall impact of a long list of potential project risks that were each assigned a relatively low likelihood. From a dozen or so risks that each have an assigned (uncorrelated) likelihood of around 10 percent, one can reasonably expect that at least one of them will be realized over the course of the project.

Tracking the evolution of project risk over time is often conveyed as a "risk burndown chart," which graphically represents the evolution of a project performance metric that is expected to diminish over time. Risk is one example; another example (used in Agile software iterations) is work remaining before some intermediate deadline. Figure 4.6 illustrates the idea. The sum of all the risk factors in the risk registry (indicated as the fifth column in table 4.3) ought to diminish as the project moves forward, with appropriate investment of contingency funds. The risk burndown chart is a way to visualize that progress. For this to be useful, the risk registry has to be kept up to date. Experienced project managers will convene regular meetings for the sole purpose of reviewing the project's risk register.

One might hope that the risk registry would contain the full inventory of all possible misfortunes that might arise on a project. There are certain to be things we miss, however—a global pandemic was likely missing from the risk registry of most major projects that were under way in 2020, for example.

There are rather sophisticated risk assessment tools that can be brought to bear on assessing whether a project has adequate remaining contingency resources, compared to the quantified risk registry. Monte Carlo simulations can be run based on the assigned risk likelihoods and consequences.

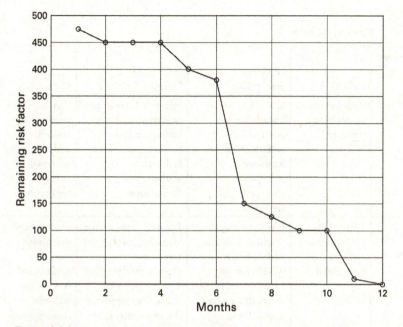

Figure 4.6
Risk burndown chart. This example shows the evolution of assessed risk for a one component of a notional project. The first successful prototype was shown to work in month 7, followed by burn-in. The first production unit passed its validation tests in month 11. It was successfully placed in service in month 12.

This can provide a histogram showing a probability distribution of possible risk outcomes, which can guide decision-making and resource allocation. It's important to remember, of course, that the outcome of that exercise is only as good as its input data, and predicting risk likelihood is rather inexact.

4.4.1 TECHNOLOGY READINESS LEVELS (TRLS)

NASA's formulation of technical maturity is a useful and widely adopted framework for assessing and managing technical risks associated with the degree of technical completion of both hardware and software. Table 4.4 is an extension of table 2.1, with the addition of columns for software readiness and criteria for transitioning to the next level. The material in table 4.4 is reproduced from ("NPR 7123.1C—Chapter1" n.d.).

The European Space Agency also uses this TRL 1–9 scale. While these TRL definitions have elements that pertain to space-qualification of flight

Table 4.4
Technology Readiness Levels

Technology Readiness Level	Definition	Hardware description	Software definition	Exit criteria
1	Basic principles observed and reported.	Scientific knowledge generated underpinning hardware technology concepts/ application.	Scientific knowledge generated underpinning basic properties of software architecture and mathematical formulation.	Peer-reviewed publication of research underlying the proposed concept/ application.
2	Technology concept and/or application formulated.	Invention begins, practical application is identified but is speculative, no experimental proof or detailed analysis is available to support the conjecture.	Practical application is identified but is speculative, no experimental proof or detailed analysis is available to support the conjecture. Basic properties of algorithms, representations, and concepts defined. Basic principles coded. Experiments performed with synthetic data.	Documented description of the application/ concept that addresses feasibility and benefit.
3	Analytical and experimental critical function and/ or characteristic proof of concept.	Analytical studies place the technology in an appropriate context and laboratory demonstrations, modeling and simulation validate analytical prediction.	Development of limited functionality to validate critical properties and predictions using non-integrated software components.	Documented analytical/ experimental results validating predictions of key parameters.

Table 4.4 (continued)

Technology Readiness Level	Definition	Hardware description	Software definition	Exit criteria
4	Component and/or breadboard validation in laboratory environment.	A low fidelity system/component breadboard is built and operated to demonstrate basic functionality and critical test environments, and associated performance predictions are defined relative to the final operating environment.	Key, functionally critical software components are integrated and functionally validated, to establish interoperability and begin architecture development. Relevant environments defined and performance in this environment predicted.	Documented test performance demonstrating agreement with analytical predictions. Documented definition of relevant environment.
5	Component and/or breadboard validation in relevant environment.	A medium fidelity system/component brassboard is built and operated to demonstrate overall performance in a simulated operational environment with realistic support elements that demonstrates overall performance in critical areas. Performance predictions are made for subsequent development phases.	End-to-end software elements implemented and interfaced with existing systems/ simulations conforming to target environment. End-to-end software system, tested in relevant environment, meeting predicted performance. Operational environment performance predicted. Prototype implementations developed.	Documented test performance demonstrating agreement with analytical predictions. Documented definition of scaling requirements.

Table 4.4 (continued)

Technology Readiness Level	Definition	Hardware description	Software definition	Exit criteria
6	System/subsystem model or prototype demonstration in an operational environment.	A high fidelity system/component prototype that adequately addresses all critical scaling issues is built and operated in a relevant environment to demonstrate operations under critical environmental conditions.	Prototype implementations of the software demonstrated on full-scale realistic problems. Partially integrate with existing hardware/software systems. Limited documentation available. Engineering feasibility fully demonstrated.	Documented test performance demonstrating agreement with analytical predictions.
7	System prototype demonstration in an operational environment.	A high fidelity engineering unit that adequately addresses all critical scaling issues is built and operated in a relevant environment to demonstrate performance in the actual operational environment and platform (ground, airborne, or space).	Prototype software exists having all key functionality available for demonstration and test. Well integrated with operational hardware/software systems demonstrating operational feasibility. Most software bugs removed. Limited documentation available.	Documented test performance demonstrating agreement with analytical predictions.

Table 4.4 (continued)

Technology Readiness Level	Definition	Hardware description	Software definition	Exit criteria
8	Actual system completed and "flight qualified" through test and demonstration.	The final product in its final configuration is successfully demonstrated through test and analysis for its intended operational environment and platform (ground, airborne, or space).	All software has been thoroughly debugged and fully integrated with all operational hardware and software systems. All user documentation, training documentation, and maintenance documentation completed. All functionality successfully demonstrated in simulated operational scenarios. Verification and Validation (V&V) completed.	Documented test performance verifying analytical predictions.
9	Actual system flight proven through successful operations.	The final product is successfully operated in an actual mission.	All software has been thoroughly debugged and fully integrated with all operational hardware/software systems. All documentation has been completed. Sustaining software engineering support is in place. System has been successfully operated in the operational environment.	Documented mission operational results.

This structure defines levels of technical maturity for hardware and software, from conceptual stages to fully validated realizations.

Table 4.5

Relationship between Critical Decision (CD) and Technology Readiness Level (TRL) for U.S. Department of Energy

Project phase	Minimum TRL for critical technology elements
CD-1, Identify Alternatives and Bracket Cost Range	4
CD-2, Set Baseline Budget and Execution Plan	6
CD-3, Begin Construction	6
CD-4, Begin Operations	7

systems, the overall philosophy has been widely adopted across the technical project management community as a way to capture and communicate the maturity of critical project technical elements. The TRL level of the subsystems and components has a direct bearing on both risk management and contingency allocation.

Scientists tend to focus their attention on the first four or so TRL levels. Once something has been shown to work, even only in a rather crude fashion, so that the main issues of principle have been overcome then we consider the problem to be solved and our attention moves to the next thing. Advancing up the TRL chain, however, requires considerable time, effort, and diligence. Scientists routinely underbudget both the cost and time needed to achieve a fully mature, tested, and documented solution, be it hardware or software. Note the contention (Mankins 2009) that the cost of passing through TRL levels 7–9 can often cost far more than all the preceding levels combined.

Technical risk is implicit in the TRL formulation—the more mature a technology becomes, the less risk to cost and schedule is associated. The U.S. Department of Energy has established a rough mapping between a project's review stages and the TRL level the agency expects for technologies that are critical to project success. These are shown in table 4.5.

4.5   PLANNING FOR THE UNEXPECTED—CONTINGENCIES

While we might not be able to predict with certainly which gremlin(s) in the risk registry might spring to life and threaten the success of the project, we can be pretty sure some nasty surprise awaits. Holding some unallocated project resources (both money and schedule) in reserve as "contingency" is a standard way to handle these surprises. At the outset of a project, when cost and schedule estimates have relatively large uncertainties and there are a significant number of unmitigated risks, as much as 30 percent of the budget might be held as contingency. In addition to holding money in reserve, early in the project there is "schedule contingency" added as well, to account for the inevitable delays in getting things done. As the project progresses, these financial and schedule contingency resources are allocated in accord with both project status and risk assessment.

It's important to allocate both funds and schedule to this contingency resource. If the project has a delivery deadline, the padding between the anticipated date of completion and the delivery deadline is called "schedule float." Extending the duration of the project usually has a significant cost impact, since the people working on the project have to be paid for more time than was anticipated. The project contingency is normally held at the Project Manager level, rather than pre-allocating it across the different subsystems.

Once the baseline execution plan for the project has been established and federal funds appropriated, that sets a cost cap that is supposed to be rock-solid. Quoting from the NSF's FY 2021 budget request:

Since FY 2009, major research facility projects funded through the MREFC account have subject to NSF's "no cost overrun" policy. As a result, NSF processes and procedures must assure the development of realistic and well-supported total project cost estimates such that approved budgets for the award recipient are sufficient to accomplish the scientific objectives. The current policy as published in NSF's Major Facilities Guide (MFG) requires that: (1) the total project cost estimate when exiting the preliminary design phase includes adequate contingency to cover foreseeable risks manageable by the recipient; (2) any cost increases not covered by contingency be accommodated first by reductions in scope with any significant scope reductions reviewed by the agency prior to implementation; and (3) if the project is approved to continue and further scope reductions become too detrimental to science, then the first 10 percent of any cost increase must be covered by the sponsoring directorate through R&RA funding. NSF holds the risk to total project cost for events that are beyond the recipient's control.

This indicates the determination of the NSF to control cost growth in major projects, by stipulating that projects identify in advance the aspects they would descope, should they experience a significant cost or schedule variance. This same hardline approach is taken by DOE and NASA.

The management reserve or contingency, both funding and schedule, should be allocated over the course of project execution to either proactively retire risk or to handle the inevitable technical surprises, mishaps, and delays that arise. This is often left to the discretion of the Project Manager, in consultation with the rest of the management team.

The goal is to optimally spend this reserve down to zero by the end of the construction phase of the project. Leftover funding is a problem that is seldom encountered in large projects.

4.6   WORKING TOGETHER—COLLABORATION TOOLS

This is an area where substantial change has occurred over the past decade, and COVID-19 has also precipitated a phase transition in the nature of collaborative and dispersed teamwork. Tools that facilitate communication within a distributed team are now part of the common vernacular. Zoom, Skype, Slack, and their siblings are as useful for big science projects as they are for other purposes.

GitHub is currently the go-to standard for version control and shared development of software. If you're not already familiar with that utility and have any association with building or using software, you ought to take a weekend and figure out how it works and how you might productively use it.

4.7   HEALTH, SAFETY, AND ENVIRONMENTAL CONSIDERATIONS

This section pertains to how large construction projects contend with health, safety, and environmental issues. Some safety issues threaten the well-being of people, while others can jeopardize project success if a key component or device is damaged. Managing and minimizing adverse environmental impacts, and also minimizing the project's susceptibility to environmental threats such as earthquakes and adverse weather events, are additional important ingredients in a project's safety management plan. The management and mitigation of technical risks is a central aspect of project management and is discussed elsewhere.

4.7.1   PROTECTING HEALTH AND PERSONNEL SAFETY

Large scientific construction projects that involve major civil and structural engineering, especially in adverse and atypical environments, pose challenges to ensuring a safe and effective workplace. Safety considerations for scientific projects might also involve chemicals, potentially harmful radiation of various kinds, biohazards, high voltages, and other factors. The project management team will include one or more project safety officers who are responsible for identifying and addressing elements that might pose a health risk to personnel. A layered suite of protocols, procedures, safety and protective devices, and training requirements will be implemented to minimize hazards and risks. These processes are captured in a Project Safety Plan. One best-practice is to establish a culture in which any team member is empowered and encouraged to call an immediate stop-work declaration if they see anything they deem unsafe. Project-specific training will supplement the laboratory safety training you receive at your home institution. You should also expect to comply with site-specific safety expectations when you visit other project partner locations.

As a scientist engaged in a large project, it is essential that you understand, respect, and adhere to all safety protocols. No matter what time pressures you might encounter, there is no justification for circumventing project safety procedures.

There are ample examples of injuries and even fatalities that have occurred on scientific projects, in both the construction and operational phases. Some of these are vehicle accidents, perhaps in some cases due to overly tired individuals driving in remote locations. Other examples arise when even the best-designed safety processes fail to anticipate and defeat all possibly hazardous situations. In other instances, personnel-safety protocols were either intentionally or inadvertently ignored or circumvented.

The U.S. government tracks worker injury statistics. Averaged across all occupations, in 2018 there were 3.5 fatalities and 900 nonfatal injuries suffered for every 100,000 worker-years. The nonfatal average injury rate of roughly 1 per 100 worker-years implies that any scientific project that involves tens of people working for many years runs a significant statistical risk of an injury of some kind. We need to take this seriously and take all appropriate steps to minimize adverse health effects. If the nature of the work undertaken in your project imposes a risk above the national average, this likelihood is increased accordingly. So should your vigilance.

4.7.2   ENVIRONMENTAL, CULTURAL, AND ETHICAL CONSIDERATIONS

The interaction between a science project and the environment goes both ways, ranging from the carbon footprint of convening in-person team meetings and reviews to the risks posed to facilities by wildfires, floods, and earthquakes. Responsible and ethical project management considers and addresses all of these factors. The site impact of a new facility is one consideration when federal agencies evaluate whether to embark a new proposed project.

The cultural, social, and ethical aspects and ramifications of a proposed project all merit attention as well.

4.8   CYBERSECURITY CONSIDERATIONS

We need to acknowledge and contend with the fact that our computer systems are constantly subjected to cyberattacks. Project cybersecurity concerns range from disruption and harassment to ransomware attacks to phishing attempts and intrusion into hardware control systems, databases, and code repositories. If the project involves export-controlled or privacy-controlled information then there are potential legal ramifications to successful cyberattacks.

A sobering cyber event that impacted a major science facility began on October 29, 2022. The ALMA international radio telescope facility in Chile suffered a cyberattack that took the entire observatory offline for forty-eight days. With an annual operating cost of around $90 million/year, each day the facility is offline incurs a direct cost of around $250,000 per day, with no scientific return. This operations cost figure does not include the impact on research and education across the community of ALMA users. The ALMA experience is a stark reminder that cybersecurity is a very real issue and merits our full attention.

Widely dispersed scientific collaborations that use common electronic tools for information, code, and data exchange are only as secure as their weakest link. Achieving an appropriate balance between scientific and technical productivity and cybersecurity is a nontrivial IT and management challenge.

Across all aspects of your professional life (and yes, your personal life as well!) it's essential that you adhere to cyber-hygiene best practices. Never, ever, share or post usernames and passwords. Keep operating systems up to date with the most recent patches and updates. Have patience with the IT professionals who are contending with very real and often very sophisticated hacking attempts.

There is an analogy here to personnel safety considerations. The hazards in both health/safety and cybersecurity are very real, and have impacted projects just like yours. Take it seriously and support your teammates who are dealing with high-consequence threats and hazards.

4.9 CHANGE CONTROL MANAGEMENT

As a major construction or software development process proceeds, the project typically finds that there are desired changes compared to the design baseline. Rather than leaving to individuals the decisions about whether departing from the (costed!) baseline design is a good idea, large projects implement a well-defined process for evaluating proposed changes. The Change Control Board is the typical project management structure that evaluates and can authorize proposed departures from the baseline design.

Factors that are weighed in considering a proposed substantive change include cost, risk, schedule, and performance implications. Scientists are likely to find project managers reluctant to embark on design changes that constitute scope creep even if there are corresponding performance gains.

The change control process typically starts with a formal request that describes the rationale, pros, and cons of a proposed change. The Change Control Board has representation from project managers of all the major subsystems who can assess any ripple effects of a proposed change. The Change Control Board usually serves in an advisory capacity to the Project Manager. Representation by scientists on the Change Control Board tends to be low, for reasons that are rather obvious.

4.10 CLOSING THOUGHTS ON TECHNICAL PROJECT MANAGEMENT

If this is the first time you're encountering the framework for technical project management, it might seem both alien and daunting. If you have questions on how this methodology is being brought to bear on a project in which you're engaged, ask around and read the relevant documentation. Briefing materials and proposals are also valuable sources of information and perspective. Engage.

Setting up all this project management structure does not in any way assure project success. It provides a structure for teamwork and cooperation, and a way to steward and allocate resources. In no way is it a substitute for this teamwork.

One specific pitfall in the Work Breakdown Structure approach is that the early phases of a project tend to focus on the completion of rather independent subsystems. This siloed mindset can persist into the system integration and commissioning phases, when those different subsystems are brought together. If at that stage the project encounters difficulties there can be a tendency to engage in finger-pointing and blame among subsystems rather than a full-team focus on identifying and solving the underlying problem(s), regardless of where they reside. There are steps that be taken along the way to minimize the chances of having the project team engage in a technical version of a civil war. These include:

- Celebrate intermediate milestones and accomplishments as an entire team.
- Find early opportunities for integrated testing.
- Clarify the ways in which the systems integration and commissioning team can call upon subsystem resources to solve integration-level problems.
- Make it clear to all participants that overall system success is the priority, not having individual subsystems delivered on time and budget. The latter does not ensure the former.
- Allocate adequate contingency resources, both in funds and schedule, to allow for surprises at the system integration stage.
- Establish a project culture of social functions that include all participants, and instill a spirit of shared enterprise and excitement.

EXTERNAL REVIEWS ONCE BIG PROJECTS ARE UNDERWAY

Periodic in-depth project reviews are one area where big-science differs from the life cycle of single-investigator grants. This chapter starts with an overview of the goals, structure, and typical agenda of these in-depth reviews, followed by sections on how to prepare and present a review talk, and review etiquette. The final section of the chapter presents suggestions for individuals who are asked to serve as members of a review panel. Appendixes 1 and 2 contain a checklist for the preparation of review presentations and an example of a review committee charge, respectively.

Agency oversight of major projects is typically conducted through peer reviews, in which an external panel engages in an assessment of the technical and managerial status and progress of the endeavor. These reviews often extend across multiple days, with breakout sessions that drill down into specific technical and management aspects. Periodic "progress" reviews track how execution is progressing compared to budget and schedule. Major milestone reviews also occur at the Conceptual Design Review (CDR), Preliminary Design Review (PDR), Final Design Review (FDR), and Operational Readiness Review (ORR) stages. Each of these reviews is a major undertaking and requires considerable investment of time and effort of key project personnel for document preparation, logistics, presentation development, dry runs, and eventually running the review itself. In addition to full-project reviews, it is common for large projects to also have dedicated in-depth reviews of major subsystems. These subsystem reviews usually follow the script and structure outlined below, but with a narrower focus.

Figure 5.1 shows the peer review–evaluated decision stages that are used by the U.S. Department of Energy for major projects. These are supplemented with periodic status reviews, at least annually.

The review panel will be given an explicit "charge" from the funding agency. Sometimes this charge document is used to guide the panel's

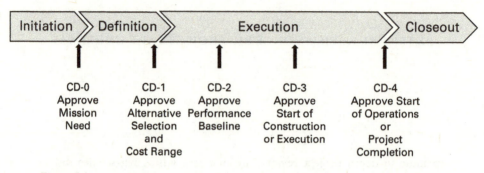

Figure 5.1
Critical Decision (CD) points for U.S. Department of Energy major projects. Each of these agency decision gates includes a substantial corresponding peer review session. Adapted from Frank 2018.

focus toward specific areas of concern. Appendix 2 provides an example. The questions posed to the review panel in the charge document are often phrased to elicit yes/no answers, such as:

• Is the project progressing as planned, and are they making appropriate plans for future construction and commissioning work?
• Is the project management functioning well?
• Are technical risks properly identified, and are appropriate risk mitigation steps being successfully pursued?
• Are Environment, Safety & Health (ES&H) issues managed appropriately?
• Has the project responded satisfactorily to recommendations from previous reviews?

The review panel will typically respond with "yes," "no," or "yes, except for . . ." answers to these top-level questions. In addition, the review panel will typically provide more detailed assessments, findings, and recommendations at a more fine-grained level. A verbal "outbrief" executive summary from the review panel is commonly provided at the end of the review session, followed by a more in-depth written report. The written report is usually delivered to the funding agencies and made available to the project team.

The membership and the chair of the review panel are typically designated by the funding agency, with an attempt to achieve the blending of expertise appropriate to the task. Ideally, the chair is someone with extensive experience in these settings, who can organize the review panel into an effective team. Agency program managers are often considered ex-officio observers and attend the review presentations.

Review materials, comprising background documentation as well as presentation overheads, are usually provided to the reviewers well in advance. This allows them time to review the materials and assemble an initial set of high-level questions as well topics for in-depth assessment. The detailed content of these review materials depends on the phase of project execution. In the initial stages a lot of attention is placed on the basis of cost estimation, and the identification and prioritization of technical risks. Later on, execution progress is tracked against budget and schedule. At the final stages of a project, the reviewer's attention shifts to demonstration of key performance parameters, the assignment of contingency resources, and the transition to operations. It is not uncommon for the review panel to be granted electronic access to the entire archive of project documentation.

5.1   TYPICAL REVIEW STRUCTURE AND AGENDA

Project reviews usually begin with an executive session for the review panel. Members of the project team do not normally attend this portion of the meeting. An experienced panel chair will use that time to review the charge and to assign specific areas of focus and writing tasks to each review panel member. The open-session kickoff talk is usually next, with a high-level overview and status report given by either the project director or the project manager. The first half of Day 1 is usually rounded out with subsystem status updates, followed by topical breakout sessions that might each only have one or two review panel members present. There is often a dinner at the end of Day 1 that provides an opportunity for additional conversation in a less formal setting. In some instances, the review panel will hold an end-of-day executive session, after which they might present a list of questions to the project team with the expectation of hearing answers the following day. This can make for a long day if team members are also expected to attend a dinner with the panel members! Plan accordingly. Depending on how things go, the review panel might also ask for last-minute changes to the agenda. Coffee breaks and meals are also an integral part of the review.

The review typically winds down with an executive session for the panel, during which they prepare a brief summary of findings and recommendations that are presented in the final "closeout" session of the review.

5.2 PREPARING AND DELIVERING A REVIEW PRESENTATION

If you're asked to prepare a presentation for a project review, you need to appreciate that many people are counting on you to invest the time and effort needed to deliver a high-quality product. You're representing the entire team. The project manager is likely to serve as the overall coordinator for the review and should communicate the structure, scope, and time allotted for your presentation. The project manager is also likely to distribute template slides and ask you to use them in order to convey a sense of cohesion and unity on the team.

Even if you're a fairly junior scientist affiliated with the project, don't be surprised if you're asked to make a review presentation. The extent to which the project is successfully cultivating a pipeline of talent is, for some agencies, an important element of the assessment process.

Appendix 1 contains a checklist for scientists who are asked to make a project review presentation.

5.2.1 CONTENT FOR REVIEW PRESENTATIONS

Preparing for these review presentations is different from preparing for an academic seminar. It's important to fully understand the main points that are expected to be communicated in your presentation, and how they fit into the broader structure of the review. Also, it's important to know what materials the reviewers will have seen already. For example, if you're giving a talk in a breakout session you likely can dispense with the introductory motivational material that might appear in a stand-alone presentation. Appendix 1 provides a summary check-list for the preparation of review presentations.

If you're a scientist embedded in a project team who is giving a talk at a review, it's likely your presentation content leans toward technical material rather than managerial, administrative, or fiscal. Nevertheless, you should link your talk's content to the project's Work Breakdown Structure and risk management register. This can be lightweight, perhaps with a reference to the WBS element on your title slide. If significant technical risk resides in your area, the project manager will likely want you to acknowledge that fact and perhaps even describe how it's being mitigated.

One of the challenges in preparing a presentation for an external review is budgeting time for the presentation, given the wide range of reviewer interaction one might encounter. The presenter might not get past a couple of overheads, and can spend the entire time answering reviewer questions. Alternatively, the reviewers might tune out and the speaker has to fill the entire time. It's hard to know in advance what will unfold during your talk. My advice, based on seeing both extremes occur, is to prepare a talk that gets to a conclusion slide in about half the allotted time if given with no interruptions. Then add stand-alone backup sections that can fill the other half of the time if needed, that delve a bit deeper into specific technical areas. If the review panel members have not taken up your allotted time with interruptions and questions, you can glide into the backup material with a transition like "Since we have additional time you might be interested to know more about how we performed system calibration . . ." If you're feeling pressed for time with an urgent desire to get to important material later in the presentation, tell the reviewer who's in charge of the session. Be sure you get the most important points across, from the perspective of the charge questions.

The backup slides serve another purpose—you can probably anticipate many and perhaps even most of the questions that reviewers are likely to raise. Having backup slides on hand that add depth and detail to those issues is very useful. Since the slides are usually distributed well in advance of the review, the backup slides can reassure your reviewers about technical issues that might not have time for discussion in a packed review agenda. Also, the backup slides can include material tailored for different levels of reviewer expertise and you can use them accordingly.

5.2.2   ADDRESS THE CHARGE QUESTIONS

It's important to be sure to give the review panel the material and information they need to address the questions in the review charge. Be sure you understand what element of the review you are expected to cover. Put yourself in the place of a review panel member who is supposed to address the review questions, and include appropriate material in your presentation.

Review panelists will want to understand to what extent the safety, cost control, and risk management protocols have been absorbed by all levels of the project team. That includes you. Be sure you know the answer to questions like "Do you understand the process for identifying and tracking a newly discovered technical risk in the project?" and "What are the main hazards associated with this aspect of the project, and what training have you received to deal with it safely?"

5.2.3   REHEARSALS FOR PROJECT REVIEWS

The run-up to the review usually includes a dry run where the project presenters rehearse their presentations in front of their teammates. Expect dry run talks to happen two weeks or more in advance of review dates, and plan accordingly. It's important that you attend dry runs for overall project overviews and for other presentations that are related to yours. This is the time to identify and resolve differences of opinion, as opposed to during the actual review. The dry run is the right time to ask questions of project team colleagues, draw attention to any inconsistencies, and find opportunities to increase clarity. Assign a notetaker colleague for your dry run talk, who can jot down comments and suggestions from other team members, and be sure that your talk will comfortably fit into the time available. Expect to make some tweaks and corrections to your talk, and budget time for that. Complete your final talk slides well in advance of the date when the materials are made available to the review panel.

You should expect that all the talks for your session, including yours, will be loaded onto a single computer, connected to the projector in the session room. This minimizes and ideally avoids the inefficiency and chaos that seems to always accompany many speakers connecting their personal laptops to the projector. Make sure that graphics and equations and fonts all work properly on the actual review computer and projector, well in advance. This is another benefit of having a review rehearsal.

The dry run might include individuals who are designated as "red team" members, whose role is to ask harsher and harder questions than the project team is likely to encounter from the actual review panel. This can be very helpful in identifying weaknesses with enough time to address them before the final version of the review materials are provided to the review panel. If this approach is adopted by your project manager, you might consider volunteering to serve as a red team member. It will give you valuable insight and perspective on the overall project.

5.2.4 STAY WITHIN YOUR ALLOTTED TIME

Under no circumstances should your presentation exceed its allotted time slot in the review agenda. A well-organized project manager will have sorted out timekeeping management for the review sessions ahead of time. Either a review panel member or someone on the project team is normally tasked with keeping the session on track and on time. If you find yourself giving a talk in a review session with no clearly designated timekeeper, then it's up to you to make sure you don't run over.

5.2.5 EXPECT TOUGH QUESTIONS

You should expect reviewers to ask tough questions, and to push back on your answers. That's why they are there. Don't take offense. The trick is to convey the sense that you know your stuff with an appropriate depth of understanding, without engaging in a bristling argument with a reviewer. If the session is straying into an extended discussion on a noncentral technical issue, you can suggest to a highly engaged reviewer that you discuss it at the next coffee break.

It's likely that other team members will be in attendance during your presentation. If a question arises for which one of your colleagues is likely to have a better answer than you, feel free to call on them to provide the response. Big projects are a team effort. It's gracious to recognize key team members who made important contributions to the content in your talk. If you don't know the answer to a question, don't try to fake it. The reviewers will know. Far better to say something like "I'll confer with my colleagues and get that answer to you by the end of the day." Don't interrupt a reviewer if they're asking a question or making a comment. Let them finish and then respond.

5.2.6 BE AGILE

Don't be surprised or offended if you find that your presentation has been eliminated in a last-minute rearrangement of the agenda. Maybe the reviewers looked though your slides ahead of time and decided they had learned what they needed from the online materials. Perhaps there is some other aspect of the project where they decided to focus their attention. If on the other hand you find that *your* piece of the project has emerged as an area of focus for the review panel, be ready to burn the midnight oil in partnership with your colleagues to address the review panel's issues of interest or concern.

Accept constructive criticism with an open mind and with grace. The reviewers are likely not as deeply obsessed with your piece of the project as you are, and that can be a useful thing. The external reviewers are often very experienced in similar undertakings, so be sure to give careful consideration to their feedback and suggestions.

5.3   REVIEW ETIQUETTE: DO'S AND DON'TS

There are some unwritten rules and expectations regarding your conduct during a review session, and I'll try to spell some of these out in this section. First of all, don't expect to be able to attend anything labeled "Executive Session" or "Panel Discussion" on the agenda. If you have any questions about what review sessions are open to you, ask the project manager or another member of the review administrative staff.

Any and all conversations with or within earshot of review panel members or agency staff should be considered as instances when you are being assessed as a team and as an individual. That includes coffee breaks, meals, and any other encounters. Be your best self.

If you're among the individuals making a presentation at the review or if you have a position of leadership on the project, it's expected that you'll show up for the plenary sessions at the beginning and end of the review process. That includes the opening overview talks and the wrap-up outbrief at the end. Turn up on time, and don't leave the room until the sessions are over. Drifting in or out of the room in the middle of a talk is disruptive and discourteous. Attending breakout sessions and supporting teammates is generally considered a good thing, subject to the considerations enumerated here. Be on time for everything, including review sessions, breaks, and meals. Be honest and ethical.

External peer review sessions are not the right time to ask presenters various questions that might interest you, or to interject some good idea that suddenly occurs to you. Make a note and find another time to discuss those issues with your teammates. Similarly, if you're attending a session for which you aren't the speaker, it's not your role to answer a reviewer's question unless you're invited to do so by the presenter. You can give the presenter a subtle hand wave to catch their attention if you think you can add to the team's response on some topic.

It's considered poor form to appeal to the reviewers to increase resource allocation to your portion of the project, either during a formal review session or during the surrounding social engagements.

5.4 SERVING EFFECTIVELY AS AN EXTERNAL REVIEWER

Let's now turn around the roles in a review. If you're called upon to serve on the review panel for a major project, I encourage you to accept. Doing so is an opportunity both to learn and to draw upon your experience and insights to contribute to the strength of the scientific enterprise. Major project reviews are a multiday time commitment. Not only is there usually a day-and-a-half or two of review presentations, but you'll usually need to do some work both beforehand to fully prepare for the review and afterwards to produce a final written report.

One tension that arises during reviews is who controls the flow, and how time is spent. While the project team may have prepared a suite of presentations, the panel is not obliged to hear them all. If time is being wasted on things everyone knows, or there is any hint of the project filling-in the review time, the panel members should assert control of the review. I've been on either side of reviews where the schedule is amended in real time and reviewers dispense with a scheduled presentation in order to pursue a topic of concern.

Reviews constitute a major investment of time both for the project and for the reviewers. Taking into account the time spent in preparation as well as the presentation schedule, a major review can easily consume person-years of equivalent effort. As a reviewer, do whatever you can to ensure that time is well-spent. If you think a particular presentation is wasting people's time, you can interrupt and suggest that the presenter focus on main topics, or else that the session shift to question-and-answer.

As a reviewer you have a responsibility and obligation to fully disclose any and all potential conflicts of interest, or anything that might be perceived as a potential conflict of interest. In addition to the rather obvious things, such as having a financial interest in any company doing business in an area close to the project, this should include collaborations with individuals on the project team or who are members of the science collaboration, as well as any ways in which you or your home institution stands to benefit from the panel deliberations.

These declarations don't usually disqualify you from review participation. You might have to recuse yourself from certain topics, but even that is unusual. If you have domain expertise, then you likely were or are engaged in a similar or competing project. What matters most is that you not compromise the integrity of the review or your professional reputation. A full and complete disclosure of any and all possible conflicts of interest will ensure that outcome. There are a variety of ways that your potential conflicts can be managed, including:

1. Partial recusal or withdrawal, so you do not receive relevant portions of the review materials, and are not present for discussions on those topics. If this approach is taken, be sure to adhere to these expectations and maintain the integrity of the review process.
2. Be present and fully informed, but in a limited role for relevant portions of the review. In this instance you might be asked to answer questions of clarification but not render any opinions or weight in on evaluations on the aspects where a perception of conflict might exist.
3. Be a full participant in all aspects of the review, with a notation that the potential conflict was evaluated and properly adjudicated.

It's important to understand the maturity expected of the project and of the project team for the review in question. Detailed implementation issues are inappropriate for a conceptual design review, for example. In a well-organized review, the specific questions the panel is expected to answer should be provided to you well in advance.

The review panel chair (usually selected by the funding agency in consultation with project leadership) will often assign certain elements of the committee's charge to specific members of the review panel, usually in alignment with their particular areas of expertise. The executive session at the start of the review is usually attended by one or more oversight officials of the funding agencies, and that's the opportunity to make sure the panel has a clear understanding of expectations, and in what form the review results should be delivered. The agency officials might also take the opportunity to highlight areas of concern, such as a technical or managerial issue that arose in a prior review.

During the time scheduled for panel deliberations you'll be expected to lead panel discussions on your assigned topics. You should also expect to provide a slide or two to any outbrief delivered to the project team at the end of the review.

In addition to contributing to one or more subsystem-level technical reviews, you have a role in reviewing the project in its entirety, including the managerial, project management, and team-cohesion aspects. That top-level assessment is carried out by the entire review panel.

The review panel is expected to provide yes-no-maybe answers to specific questions in the charge, with appropriate justification. The panel is also usually asked to make observations and recommendations for corrective actions. These corrective actions fall into must-do, should-do, and suggest-you-consider categories. One element of the review will be the extent to which the project addressed corrective actions from prior reviews.

5.4.1   DO YOUR HOMEWORK AHEAD OF TIME

Each of the main review stages has associated documentation, the volume of which increases with project maturity. Reviewers will usually be sent a link and credentials for accessing these materials, online, ahead of the review. Often, the review presentations will be posted and made available ahead of time as well.

Take full advantage of this early access to:

- Review the presentations most directly relevant to your segment of the review, and jot down questions that are triggered.
- Read the high-level documents and become acquainted with the system and the team.
- Study the technical and managerial documents that pertain to your domain of expertise or area of review focus.
- Drill down into a few detailed items such as procurement justifications, technical memoranda, and design documentation.

This top-to-bottom scan of the project materials will provide a solid foundation for your engagement at the time of the review and presentations. You may find gaps in the project execution flow or inconsistencies up and down the project structure. Alternatively, you may find a taut and consistent structure with a clear plan that is being well-executed. Bring any concerns you have to the review, and raise them during the initial executive session. This will alert other members of the review panel to these issues.

The panel can also provide questions to the project team at any stage of the review, including after the initial executive session.

If you think the posted presentations miss the mark or are not responsive to the committee's charge, let the team know.

Much (if not most) of the value in conducting reviews comes from forcing the project team to complete a well-defined checklist of preparatory tasks by a well-defined deadline. Going through the checklist associated with a given review imposes a discipline that helps the project attain a uniform level of maturity across the different tasks in the WBS. One of your tasks as a reviewer is to assess the extent to which they succeeded.

5.4.2   ENGAGE, AND ASK QUESTIONS

A review shouldn't be a series of seminars where the reviewers constitute a passive audience watching the show. Engage with the presenters. Ask questions that gauge their level of mastery of the material, and the level of coordination and cooperation within the project team. This does not require falling into a pattern of abusing your reviewer's status by providing an ongoing running commentary. A few well-posed questions can determine a great deal. If an issue is uncovered, the review is neither the time nor the place to resolve it. Spending more than five minutes on any one technical issue is likely the point of diminishing returns. Further discussions can be pursued afterwards, or in correspondence with the project team.

Be attuned to the conversations and interpersonal dynamics during the presentations as well as at breaks and meals. Do these people act as a team? Is there a sense of shared purpose? Do they understand their respective roles, responsibilities, and authorities? Do they know how to place a new entry into the project's risk registry?

5.4.3   PROVIDE CONSTRUCTIVE FEEDBACK

The review process is an opportunity to provide useful feedback to the project at three levels:

1. *The formal response to the review charge.* This is usually fairly terse and formulaic, with a few sentences of justification for each item.
2. *Recommendations for action items for the project.* Be judicious in things that are stipulated as required actions. These can be in both the technical and managerial domains.

3. *Informal technical and managerial advice and suggestions to the project team.* These are often conveyed in informal sidebar discussions, and sometimes in follow-up email correspondence.

While there is an evaluative aspect to the review process, concrete suggestions that are well-reasoned, actionable, and constructive are likely to be the most useful to the project team. With this in mind, it's useful to keep notes in these three categories over the course of the review.

ORGANIZATIONAL STRUCTURES FOR PROJECTS AND COLLABORATIONS

A scientist might be inclined to gloss over organizational charts, but it's essential that project participants, including you, understand the authorities and reporting relationships that are conveyed in these diagrams. This chapter will show a few examples that should help you decode the implicit values and priorities that are conveyed by these structures. To paraphrase a quip used by politicians, "Don't tell me what you value, show me your org chart and I'll tell you what your community values the most."

Figure 6.1 shows the notional degree of scientist/user participation in facilities construction, for different scientific subfields. The nature of the facility plays a strong role (D'Ippolito and Rüling 2019) in the choices scientists make in structuring their efforts, as does the extent of overlap shown in figure 6.1. For user facilities that serve many disjoint groups (synchrotron light sources come to mind), the scope of a science collaboration is unlikely to extend beyond a single experiment, or perhaps a beamline. Experimental particle physics, on the other hand, is an example of a field where the practicing scientists are deeply embedded in the development and operation of facilities.

In fields other than experimental particle physics, the builders/operators of facilities are typically less tightly linked to the science users. Examples that span a range of facility/user relationships include optical and radio telescopes, research vessels, synchrotron light sources, remote-sensing Earth-observing satellites, high-throughput gene sequencing facilities, nano-fabrication, and materials analysis centers. For many of these, the facilities/execution structure and the science collaboration are distinct entities.

6.1 A HIGHLY INTEGRATED SYSTEM: CERN

An example of a system where the facility is tightly integrated with the science/user community is CERN (originally the Conseil Européen pour

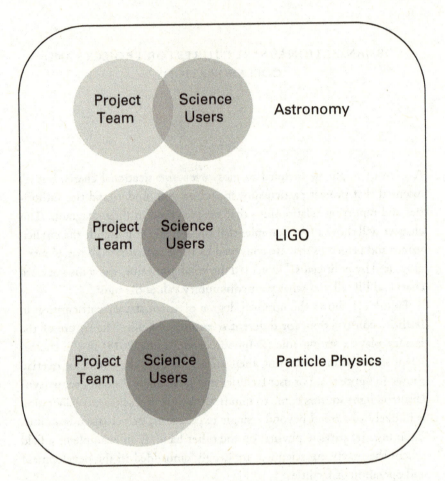

Figure 6.1
Notional extent to which the science/user community, for different subfields, is
engaged in the construction of new facilities. The degree of scientist engagement with
major projects depends heavily on subdiscipline norms and traditions. From top to
bottom, the overlap increases as one moves from astronomy facilities to the LIGO
gravitational wave experiment to collider particle physics experiments.

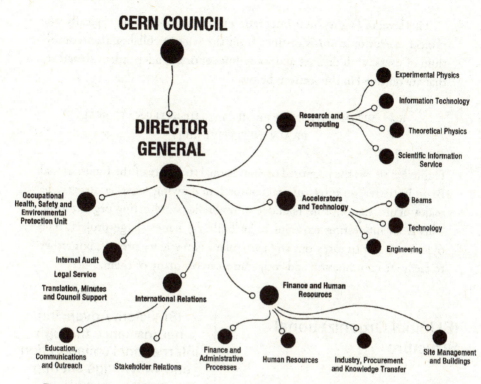

CERN COUNCIL

DIRECTOR GENERAL

Research and Computing
- Experimental Physics
- Information Technology
- Theoretical Physics
- Scientific Information Service

Occupational Health, Safety and Environmental Protection Unit

Accelerators and Technology
- Beams
- Technology
- Engineering

Internal Audit

Legal Service

Translation, Minutes and Council Support

International Relations

Finance and Human Resources

Education, Communications and Outreach

Stakeholder Relations

Finance and Administrative Processes

Human Resources

Industry, Procurement and Knowledge Transfer

Site Management and Buildings

Figure 6.2
Organizational structure of CERN, with an integration of facilities, science users, and other program elements.

la Recherche Nucléaire), arguably the largest science endeavor currently in operation. The CERN ("About CERN—CERN Environment Report 2019–2020" 2020) organizational structure is shown in figure 6.2.

The management of projects at CERN is particularly challenging, with nations contributing cash, hardware, and software to this enterprise. The complexity (technical, legal, financial, political, etc.) of this operation necessitated the early adoption and ongoing evolution of technical project management methods.

CERN's structure includes the direct incorporation of experimental and theoretical physics and facilities development/operations. Another notable feature is the inclusion of what amounts to a diplomatic corps to nurture the level of international cooperation needed to sustain the effort.

This level of facility/user integration is uncommon. More typically, the project execution team is distinct from the science collaboration/consortium of users, with distinct and sometimes entirely independent structures that are outlined in the sections below.

6.2   A FEDERATION OF PROJECTS: THE INTERNATIONAL BRAIN INITIATIVE (IBI)

Figure 6.3 shows the proposed organizational structure of the International Brain Initiative, an international consortium drawing together an array of major neuroscience projects under an umbrella coordinating organization. This is an interesting experiment in building a meta-large-project. The objective is not to carry out any particular brain science project, but rather to facilitate coordination and cooperation across many of them.

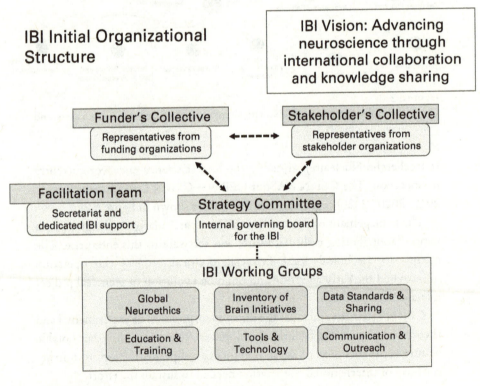

Figure 6.3
The structure of the International Brain Initiative.

The structure of the IBI shows a number of working groups, for topics ranging from setting data standards to education, as well as a consultative function to bring together both scientists and funding organizations—much support for brain science comes from philanthropic organizations. The organizational chart in figure 6.3 reflects the coordination and communication role that IBI intends to play.

6.3   A TYPICAL SCIENTIFIC CONSTRUCTION PROJECT STRUCTURE

The organizational structure for the construction of a major scientific facility blends scientists, managers, engineers, educators, and supporting administrative staff into a team that is charged with carrying out the project plan.

6.3.1   PROJECT LEADERSHIP TEAM

The Vera C. Rubin Observatory construction project's leadership structure, shown in figure 6.4, demonstrates a typical project execution organizational chart.

The Project Director, assisted in this case by a Deputy, carries overall responsibility for successfully carrying out the program. In many instances, the Project Director has a significant outward-facing role, serving as the link to federal agencies and other related projects. The Project Operations Office has a Project Manager and Project Scientist and respective deputies. It oversees and coordinates activities of four main subsystems, each with its own subsystem Project Manager and Subsystem Scientist. The systems engineering and commissioning team works to knit the subsystem elements into an integrated whole. The Project Manager holds primary authority over the engineering execution of the project plan and the allocation of project contingency. These decisions are made in close consultation with key members of the project team, but it's common for decisional authority to be vested in the Project Manager. There is often a Head or Lead Engineer, as well as a czar for the software aspects across the whole project.

Important elements of the project execution structure are the safety, compliance, and security functions. Contemporary technical project management and funding agencies have zero tolerance for anything that places project hardware or personnel at risk of harm. Safety, security, and compliance assessments are an integral part of all external reviews.

Chapter 6

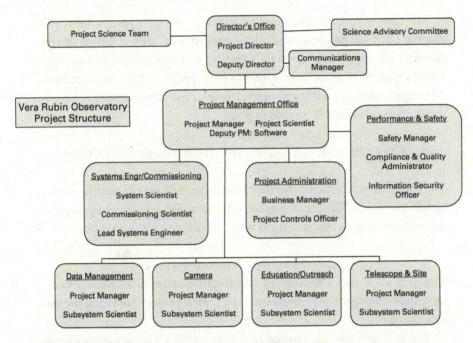

Figure 6.4
Organizational chart for the Vera C. Rubin Observatory construction project. Note
the attempt to generate creative tension by placing both a manager and scientist in
shared leadership roles at the subsystem level (bottom row). The Science Advisory
Committee is an external body, while the Project Science Team in an internal group.
Note the inclusion of explicit safety and cybersecurity roles.

The communications team provides both external and internal communities with important updates and serves to connect the various components of the project together. Administrative and IT support are provided
across the project by the cognizant support offices.

6.3.2   PROJECT ADVISORY AND OVERSIGHT STRUCTURE

The project advisory structure often consists of an internal Project Science Team and an external Science Advisory Committee, both acting in
an advisory capacity to the Project Director. This tandem advisory structure, with both internal and external oversight committees, is a commonly
used project accountability scheme. The external committee is made up
of stakeholders who have no role in project execution and is often jointly
appointed by the Project Director and the funding agencies.

Review teams are assembled by the funding agencies, usually in consultation with the project leadership team, as appropriate. They are usually disbanded after the review is finished. The funding agency will designate a program officer who will track progress on an ongoing basis.

One feature of the organizational chart in figure 6.4 is worth emphasizing. The Vera C. Rubin Observatory's construction project subsystem management scheme is somewhat unusual—the two-in-a-box combination of a subsystem Project Manager and a companion Subsystem Scientist was an explicit attempt to engineer a constructive creative tension between performance, cost, and schedule at the subsystem level.

6.3.3   PROJECT SCIENTIST

The role of Project Scientist is of particular importance to the successful execution of these projects. That individual is expected to have an end-to-end understanding of the technical aspects that are critical to the scientific success of the project. If and when crunch time comes and descoping options are under consideration, the Project Scientist has considerable influence over making choices that retain the most critical aspects of system performance.

6.3.4   SUBSYSTEM EXECUTION TEAMS

The major subsystems typically will each have an internal structure that is a smaller version of the project-wide organization. A subsystem manager bears responsibility for having the subsystem delivered on time and on budget, and reports to the Project Manager. A team of engineers and scientists work together toward that goal. There is sometimes a Subsystem Scientist designated as well.

6.4   SCIENCE COLLABORATIONS AND CONSORTIA

The project management structure put in place to execute the construction project is often complemented by a collaborative framework established by the relevant science community. For small- or mid-scale projects/experiments, the scientists who contribute to the construction are often the ones who use this collaborative framework. But as projects grow in scale, take longer, and need more money to come to fruition, and serve a larger user community, the science builders and the eventual users can become disjointed communities.

Most big-science projects are international, and this brings an additional layer of complexity, especially if the funding is being provided by multiple countries. See *Doing Global Science: A Guide to Responsible Conduct in the Global Research Enterprise* (InterAcademy Partnership 2016) for a discussion of global science collaborations.

The NSF makes a distinction ("MRI Frequently Asked Questions (Nsf15012) | NSF—National Science Foundation" n.d.) between a collaboration and a consortium:

> A collaborative proposal represents a funding mechanism, used NSF-wide, by which investigators from two or more organizations who wish to collaborate on a unified project may submit proposals and share funding. Collaborative proposals may be submitted to NSF in one of two methods: as a single proposal, in which a single award is being requested (with subawards administered by the lead organization); or by simultaneous submission of proposals from different organizations, with each organization requesting a separate award (Chapter II.D.4 of the Grant Proposal Guide provides additional information on collaborative proposals). For acquisition proposals from Ph.D.-granting or non-degree-granting institutions, only the latter method should be used. The MRI program defines consortia as a category of submission-eligible organizations to better facilitate proposals that provide access to unique instrumentation for a broad user base of U.S. scientists and engineers, and that encourage greater collaboration and sharing of state-of-the-art instrumentation (the MRI-specific definitions of consortia are included in the solicitation). The MRI program accepts proposals from two types of consortia—legally incorporated consortia located in the US with 501(c)(3) status and consortia organized specifically to support an MRI project. If a consortium has 501(c)(3) status and has the appropriate administrative and financial infrastructure to receive and manage a federal grant directly, then it may submit a consortium proposal directly. Otherwise, a consortium must include at least two MRI submission-eligible organizations, and a proposal from such an MRI consortium must be submitted by one of its submission-eligible organizations on behalf of the consortium. In either case, the proposal will count against the submitting organization's MRI submission limit. Collaborative proposals need not be consortium proposals and consortium proposals may or may not be collaborative proposals.

A collaboration typically comprises a group of scientists. The participating entities for a consortium are institutions, often bound together through a nonprofit, tax-exempt "501(c)(3)" corporation. Examples of consortia include the Giant Magellan Telescope and Thirty Meter Telescope projects, both of which have extensive non-U.S. consortium members.

The governance structure of a project can evolve over its lifetime. The Rubin Observatory began life as a consortium called the Large Synoptic Survey Telescope Corporation (LSSTC) for which the founding members were the University of Arizona, the University of Washington, the National Optical Astronomical Observatory (NOAO), and the Research Corporation, a philanthropic foundation. That consortium submitted a design and development proposal to the National Science Foundation, which funded much of the early design work on the project. This led to the submission of a construction proposal by LSSTC. But the NSF concluded that the LSSTC shell corporation did not have the track record, oversight capacity, and heft to execute a major construction project. A rapid reconfiguration by the community followed and the Association of Universities for Research in Astronomy (AURA), a long-established consortium already successfully running a number of large astronomical facilities, stepped up to take on the role of executing the NSF-funded aspect of project construction.

The scientific collaborative structures described below are distinct from external oversight groups such as a User's Committee. Our interest here is exploring the sociology of a potentially large group of scientists working on a hard problem together.

6.4.1 STRUCTURE AND GOVERNANCE OF A SCIENTIFIC COLLABORATION

There are different ways in which scientists establish governance and self-imposed structure. One well-established way to contend with the herding-of-cats problem is the benevolent scientific dictatorship. In this model one individual holds all the authority and makes decisions on behalf of the collective. In this system there tends to be a rather flat management structure below the dictator, often with significant micromanagement. Mirroring what we see around the world for nation-states, these scientific dictatorships still survive in some corners but are being displaced by more participatory forms of governance.

Figure 6.5 illustrates the organization of the CMB-S4 project. This group proposes to construct, operate, and use an ambitious system to measure the structure of the cosmic microwave background (CMB), to address outstanding issues in basic physics and cosmology. The collaboration has one

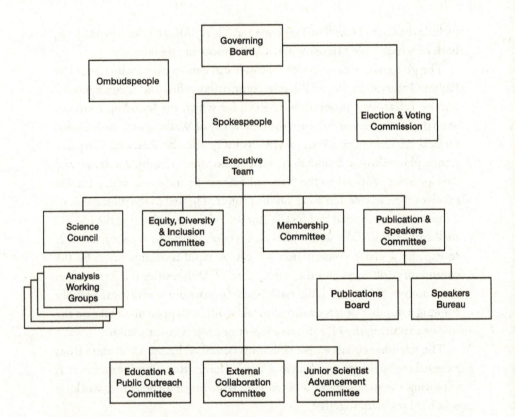

Figure 6.5
Collaboration structure for the CMB S4 project, a next-generation cosmic microwave
background project. This undertaking exemplifies a science community deeply
engaged in the facility's construction. Note the oversight from the Department of
Energy on the left. The Spokespeople (elected as prescribed in the collaboration's
bylaws) coordinate the efforts of the team. The Science and Technical efforts each
have their own management councils. The importance the collaboration attaches to
junior scientists is evident from this chart.

branch focused on technical aspects (designing and constructing the observ-
ing system) and one branch preparing for the analysis of data. These are
supplemented with one committee that grapples with membership issues,
and another that manages publications and organizing speakers for talks.
The explicit inclusion of a committee charged with nurturing the careers
of junior members of the collaboration and the designation of specific individuals as ombudspeople speak to the importance the CMB-S4 team gives
to mentoring, and to resolving instances of friction that might arise.

The Governing Board is elected from the collaboration membership, with representation of the postdoc community on this board enshrined in the CMB-S4 bylaws. This democratic approach is increasingly common in science collaborations. Two elected co-spokespeople are charged with the leadership and coordination of the enterprise.

6.4.2   MEMBERSHIP

Membership in these collaborations is usually by application, with consideration overseen by a membership committee. Scientists normally have to make a commitment of spending a certain minimum amount of their research time advancing the goals of the collaboration in order to be approved for membership. In some cases institutions rather than individuals are the "members," with that right extending to employees of the organization. There are often specific categories of membership for students and postdocs, that recognize the distinctive features of those career stages.

6.4.3   CODES OF CONDUCT

The science community is (at last) taking explicit steps to articulate and enforce expectations for professional conduct. Agreeing to comply with these shared expectations is a condition for being a member of the collaboration community. A code of conduct is useful for setting the ground rules for behavior in meetings, for relationships with power asymmetry, and for disagreements being discussed in a tone of mutual respect.

6.4.4   ENHANCING EQUITY, DIVERSITY, INCLUSION, AND BELONGING

The scientific enterprise faces substantial challenges in both attracting and retaining participants from historically under-represented communities. The education and outreach segment of the collaboration can play a vital role in establishing and enhancing relationships with relevant universities, professional societies, and individuals, but we all share responsibility for implementing lasting change.

Contemporary science collaborations bear their share of responsibility for enhancing diversity, inclusion, and belonging for underrepresented scientists. These are distinct stages along the road to inclusive excellence. Diversity refers to improving demographic metrics. Inclusion is achieved when historically underrepresented groups are participating at all levels across the collaboration structure. Belonging is achieved when each and every individual feels they belong as an integral member of the team.

Science collaborations can help at all stages of the professional pipeline. Summer internships for undergraduates, rotations through universities, national laboratories, and remote sites can broaden the experiences for graduate students. Postdocs can be attracted from other fields into an attentive mentoring culture within the collaboration. Partnerships with minority-serving institutions can enhance the career opportunities and mobility for the people they serve.

Attending to each stage of the professional pipeline, options include:

- Undergraduates—research experience, outreach
- Graduate students—full participation, membership category, protection
- Postdocs and junior scientists—nurture leadership roles
- Senior scientists—establish a culture of ownership of EBID enhancement

6.4.5 PUBLICATION POLICIES—DECLARING SCIENCE PAPERS

In order to avoid replication of effort, to forestall collisions between factions, and to protect the interests of junior members of the collaboration, it's common for collaborations to require that papers that fall within the collaboration's scope be declared at an early stage. Normally, any member of the collaboration can volunteer to contribute to any declared paper.

Collaboration members typically organize and structure their efforts through joining one or more topical working groups, each of which can have an internal governance structure with a "convener" or workgroup leader/coordinator. Work on science papers is coordinated by these individuals. Larger collaborations commonly set up an internal review process and timeline for comments prior to a paper being submitted.

6.4.6 AUTHORSHIP AND CREDIT FOR WORK DONE

One of the main advantages of participating in a well-structured collaboration is that the members agree to abide by the collaboration's authorship policy. If well-executed, the authorship policy should prevent endless discussion and anguish (Stubbs 1997) about who is listed on what paper, and in what order. In order to achieve this, the collaboration must adopt and implement a scheme that stipulates what level(s) of contribution warrant inclusion on what kinds of papers, how authors will be ordered, and the duration and portability of authorship and data rights in the event members change institutions. The authorship policies should conform to the norms and expectations of modern science and the ethical statements that have been promulgated by the various professional societies (see, e.g., *On Being a Scientist: A Guide to Responsible Conduct in Research* [National Academy of Sciences 2009]).

Authorship on papers is one of the ways the technical community keeps score. The two problematic issues are (1) who is listed as an author, and (2) the order in which their names appear. There is a branch of scholarship concerned with the study of authorship policies, and their consequences (see, for example, Patel et al. 2019; Ilakovac et al. 2007; and Marušić, Bošnjak, and Jerončić 2011). There are clearly articulated professional standards (National Academy of Sciences 2009) for determining whether an individual's contributions merit inclusion. The omission of deserving contributors and/or the unjustified inclusion of honorary authors are both inappropriate. One recent paper (McNutt et al. 2018) declares the following:

> Each author is expected to have made substantial contributions to the conception OR design of the work; OR the acquisition, analysis, OR interpretation of data; OR have drafted the work or substantively revised it; AND to have approved the submitted version (and any substantially modified version that involves the author's contribution to the study); AND to have agreed both to be personally accountable for the author's own contributions and to ensure that questions related to the accuracy or integrity of any part of the work, even ones in which the author was not personally involved, are appropriately investigated, resolved, and the resolution documented in the literature.

A common perspective is that the scientists who made substantial contributions to the system/facility that made the paper possible have earned the right of authorship just as much as the people who did the analysis and wrote the paper. How far does this extend? If certain scientists contributed essential ingredients to the conceptual design and made other important and creative intellectual contributions, most would agree to including them as coauthors. What about an engineer who was paid on project funds to do a specific well-defined circuit design? What about the people in the fiscal and accounting office? It's well-nigh impossible to craft legislation that addresses all cases that might arise and so the collaboration needs to establish a method for resolving them, based on guiding principles.

Unless the collaboration's authorship criteria are spelled out and agreed upon there is ample opportunity for misunderstanding and bitterness. Well-structured collaborations not only have an explicit authorship policy, they also establish a mechanism for the arbitration of authorship disputes.

Authorship order is the other potential source of tension between colleagues. In many disciplines the authorship order reflects relative level of contribution to a paper. There is an inherent problem, however, with any attempt to rank-order the level of contribution of a large and diverse team on a single axis, and order their authorship accordingly. This would require constructing a contribution-merit function that allocates relative weights to initial creative insight, engineering accomplishments, hardware contributions, software contributions, algorithmic innovation, and so forth, and then allocates appropriate credit to each individual.

These authorship-ordering options include:

- Rank-ordered by level of contribution as determined by some person or process
- First-authorship by "lead" person(s) on the paper, with the rest alphabetical (many journals allow designation of co-first-authorship)
- Anchor-authorship (last name in the list) by most senior person (common in life sciences)
- Alphabetical authorship order, in recognition of the difficulty inherent in any contribution-ordered scheme
- Tiered ordering, first by institutional affiliation then by contribution level
- Two-tiered ordering with lead authors for the paper in one group, and the others in a second grouping
- Authorship order determined by random number generator (Seriously! See Ray and Robson 2018.)

Each of these options has drawbacks, but ultimately some scheme must be adopted. Arriving at a solution is particularly difficult for blended teams that come into the project from disciplines with disparate expectations and traditions. Helping a fledgling collaboration through the trauma of establishing an authorship policy is one of the more delicate stages of scientific leadership.

The experimental particle physics community, which can have papers with more than five thousand authors, long ago adopted alphabetical author order. Under such circumstances authorship and authorship order become less useful as indicators of contributions by individuals—more on this below.

At the time of this writing the community is wrestling with whether to include AI entities in authorship lists (Transformer et al. 2020; Samuelson 2020), and how to handle AI-assisted manuscript preparation. Currently some journals consider the submission of AI-generated text as scientific misconduct. Other journals require simply the acknowledgement if AI text generation or editing was used. Make sure you stay abreast of evolving expectations in this area.

6.4.7 BUILDERS

The concept of a "Builder" is meant to at least partially address the scientific opportunity cost incurred by investing substantial time and effort in foundational technical work for a major project, rather than writing scientific papers. Once a scientist has invested effort into the project beyond some threshold, in some collaborations they are awarded "Builder" status, which confers a preferred category of authorship compared to other collaboration members. The justification for authorship is their essential contribution to the facility.

The builder concept has limitations. In cases where the data from the facility become public immediately, scientists outside the collaboration have no obligation to include builders on their authorship list. Also, where does one draw the line? Are all engineers and programmers who work on the project expecting to receive this designation, or is it limited to people intending to pursue an academic career?

6.4.8 RECOGNITION BEYOND AUTHORSHIP

Large collaborative efforts can produce papers that have tens or hundreds or even thousands of authors. Yet somehow people who work on these projects manage to get hired, promoted, and respected for their contributions and expertise. How does "the system" select these individuals, and figure out who did what? It is done through nonauthorship indicators of competence, leadership, and accomplishment. Ways in which individuals demonstrate these traits include:

- Taking responsibility for a specific area (for construction projects) or science objective (for a collaboration). Be the person who makes something happen.
- Holding positions of responsibility and authority in the organizational chart of the project or collaboration, whether elected or appointed
- Demonstrating mastery and proficiency as indicated by talks at meetings, seminars, summer schools, etc.
- Securing research funding through peer review
- Successfully mentoring junior colleagues
- Impressing people with work done as a reviewer of other projects

Assessments provided by collaborators, both formally through letters of recommendation and informally through professional networking, are another mechanism for allocation of intellectual credit.

Successful scientific collaborations therefore pay attention to striving for equity in who represents the effort at conferences, who gets to announce new results at major meetings, and how individuals are selected for leadership positions within the collaboration.

6.4.9   DATA RIGHTS

A critical decision for any large project is how to manage the dissemination of results and make data available. A data management plan is an essential ingredient of any serious large-project proposal. This is another area where subfield traditions vary. Individuals who have invested years, and sometimes decades, of their professional life giving birth to a new facility can feel possessive about the data that flow from the system. People outside the project can feel that resources that poured into the facility extracted a toll on other endeavors, and that the data ought to be made promptly available to all.

Particle physics experimental groups in general don't release collision event data outside the collaboration. Data collected from the Hubble Space Telescope has a one-year proprietary access period after which they become public. The Vera C. Rubin Observatory project established a policy of immediate access for data-entitled scientists. There is a clear trend toward reducing proprietary data periods and increasing ease of access to data produced by taxpayer-funded facilities, projects, and programs.

6.4.10 FUNDING IMPLICATIONS

Science work carried out by a collaboration is typically funded through federal research grants and/or philanthropic contributions. Collaboration members might also receive funding that allows them to participate in the project's construction aspects.

Funding for scientists can be allocated to the collaboration for distribution and management through the collaboration governance structure. More commonly, it is awarded directly to institutional PIs who bring those resources to bear on the project's goals. This often means that the PI's funding proposals need to demonstrate a strong link to the science objectives of the agency, as carried out by the collaboration.

This can lead to a certain amount of jockeying for meaningful leadership roles by midcareer and senior scientists in the collaboration. These are often elected positions, and so interpersonal relationships and scientific politics play a role in determining who sits where in the collaboration's org chart. It's helpful on a grant application to claim that one is playing a central leadership role.

6.4.11 CAREER PROMOTION AND MENTORING

Contemporary science collaborations take responsibility for mentoring the career progression of participants. This should augment, not replace, the advising function within each research group at their home institution. For subfields where participation in large projects is the norm, senior members of the collaboration should expect to be asked to provide letters of recommendation for junior colleagues.

In addition to clearly defining roles and responsibilities on projects, many science collaborations arrange to have members who are at a transitional career stage be the ones to represent the collaboration at conferences and workshops. This gives them a chance to "take the stage" and demonstrate their depth of contribution to and knowledge of the collaboration's science program. Since speaking invitations tend to flow to the more senior and well-established members of a collaboration, a so-called speaker's bureau is often established. Speaking invitations are forwarded to this committee, and they work to have this element of scientific recognition be fairly apportioned across the collaboration.

BEING AN EFFECTIVE MEMBER OF A BIG ENDEAVOR

Whether you are participating in and contributing to a large science construction project or are a member of a large science collaboration (or some combination of the two), the technical complexity and sheer number of participants in a big-science enterprise can pose challenges and opportunities that just don't occur in smaller-team science. This chapter provides some suggestions that might prove useful in enhancing your effectiveness in large-scale collaborative efforts. The subsequent chapter is in the same vein, but focused on junior scientists.

Much of what follows here pertains to managing interactions with colleagues. These big projects are being undertaken by a group of people, often from a wide range of personal and technical backgrounds, who need to function as a large team. The effectiveness of the interpersonal inner workings of the team directly impacts its ability to achieve its objectives, and that leads to our first topic.

7.1 CLARIFY OBJECTIVES—YOURS AND THEIRS

Being effective requires first knowing what you're striving to be effective at doing. Presumably, the big-science program has clarity of objective, be it the construction of a facility and/or the execution of a project or experiment. It is advisable, however, to give some thought as to how the big-science objectives align with your own goals, both current and longer term. Consider not only the alignment of technical and scientific objectives but also your personal and professional ambitions. Setting explicit personal goals and a rough timeline for meeting these objectives brings something of a project management perspective to your professional development, technical contributions, and science aspirations. Identify the areas in which the project's organizational structure and goals overlap with your own. Those are areas where your efforts are likely to have the maximum combined positive impact.

7.2   MINIMIZE AND HELP DEFUSE INTERPERSONAL CONFLICTS

Interpersonal conflicts and tensions seem to be an inevitable feature of any extended interaction between more than a handful of individuals. Large collaborations of scientists are no exception to this. Differences of opinion naturally arise. Career competition leads to friction. Factions and cliques develop and maneuver for influence. Often the biggest challenges that confront a large collaboration working on a hard problem are not technical in nature, they're human factors. Establishing and contributing to a sense of shared purpose can help forestall and head off potential sources in interpersonal conflict. Recognize and express appreciation for the work that others do and acknowledge their contributions.

Having the knack for resolving interpersonal conflicts is in my view what separates good scientific leaders from truly great ones. Leaders who can keep a team focused on the project objectives and minimize the time, effort, emotion, and energy expended on fussy distractions are better able to lead that team to shared scientific and technical successes. If you have any aspirations of moving up the scientific leadership ladder, you should recognize that these soft skills become increasing important at higher levels of responsibility and authority.

Find ways to cultivate and sustain a sense of full-team identity. Celebrating intermediate milestones is one good way to accomplish this. On a project that can take years or even decades to reach its final conclusion it's important to pause and recognize progress along the way. Did the main instrument pass some performance milestone? Have a picnic to celebrate. Was there a successful subsystem review? Bring in a cake. Set an example by treating all with respect and courtesy. Intervene if you witness examples of interactions that fall short of those goals. Your project will benefit from proactive actions you take.

One way to have a positive influence on team interaction dynamics is to identify trouble brewing early on and take action to diffuse it. Regardless of your level of career seniority, try to play a constructive role in understanding and addressing the underlying root causes of friction. Is it an unresolved technical disagreement? Is someone feeling slighted about a publication? Many disagreements arise from issues surrounding authorship, and a subsection below is dedicated to this topic. Another common source of (sometimes passionate) discord comes from disagreements on technical issues. This also merits a dedicated subsection below.

There are two ways to address and resolve interpersonal conflicts that arise within a technical/scientific team—informal and formal. In my experience it is far better to take action early and informally rather than letting things rise to the level where formal processes need to be invoked. This requires a willingness to recognize and contend with awkward and uncomfortable situations. But taking no action is more likely to result in escalation rather than spontaneous resolution.

If you find a growing tension between you and one or more teammates, the first step is to perform a careful self-examination. While it's easier to place all the blame on others, take a hard look at your own conduct and how others could interpret (or perhaps misinterpret) it. Try to identify the underlying source(s) of discord. Assess what actions on your part could have triggered or contributed to an awkward situation. Then try to visualize the circumstances from the perspective of others. Ask around and see how non-participants (noncombatants?) view the situation. After that attempt at seeing things objectively, initiate a nonconfrontational conversation that acknowledges the tension, and express a desire to address and resolve the growing tension. The sooner this happens, the better. Unaddressed tensions tend to fester and grow, making them increasingly harder to resolve amicably.

Email exchanges can rapidly escalate into conflict if you're not careful. If someone does something or sends a message that increases your blood pressure, go ahead and draft a blistering email response through gritted teeth (without that person's email address in the TO bar. Mistakes do happen, after all . . .). Then delete that message, cool down, and draft a professional and nonconfrontational second reply that is worthy of sending above your signature.

If you observe tensions arising that don't involve you directly, you should still feel a sense of responsibility to do what you can to help resolve them. There are pragmatic political limitations to this, admittedly. If the project director and project manager get into a heated shouting match in public and you're a postdoc who just joined the team, it's probably not a good idea to try to serve as on-the-spot mediator. But in many cases, you will likely be in a position to contribute to an informal resolution. Go have coffee, separately, with the aggrieved parties. Convey information (while respecting confidences) to people in the hierarchy who have leverage in the situation. Think about what compromises might work and suggest them. If appropriate, you can remind people about codes of conduct and team expectations of collegiality. Rally others who command respect on the team to join you in trying to resolve the issue in an informal fashion, or at a minimum ask their advice. Above all, be supportive, sympathetic, and a good listener.

If, despite all these efforts, the disagreement stubbornly resists informal resolution then the only recourse is to invoke formal processes within the collaboration. Ideally, there is a well-defined written dispute resolution process that might involve an ombudsperson or executive committee or benevolent-dictator team overlord. Setting these processes up in advance is time-consuming and rather painful, but if ever needed they're indispensable and speed the resolution of thorny issues so that people can get back to work on the project's main objectives.

7.3   HELP RESOLVE TECHNICAL DISPUTES

Disagreements on technical matters are in general easier to resolve than interpersonal conflicts that arise from differences of opinion. You can focus on the technical issue at hand, and solicit views and opinions from around the table, be that table real or virtual. A hard-nosed engineering and project management approach will serve you well here. Exactly what is the well-defined issue? What combination of design studies, prototyping, and simulations can inform a decision, and where in the project timeline must that decision be taken? What are the consequences in cost, schedule, risk, and performance associated with the various options?

Even though this is pretty standard technical project management fodder, it's important to acknowledge that people's egos and self-esteem are entangled with the stances they take on technical issues. To the extent you can help frame an unresolved technical issue as a shared team problem to be solved, you can avoid the fallout of reduced morale and a sense of winners and losers.

The trick is to maintain a sense of positive tension within the team to advance their creativity and innovation to arrive at a good solution, within a concrete project management framework and clarity of criteria for down-selection to one path forward. Strive to avoid a culture of "You won this one, so I should get the next victory."

7.4 HELP RESOLVE AUTHORSHIP DISAGREEMENTS

Conflict over authorship issues is arguably the most awkward and painful kind of interpersonal friction that can afflict collaborative science. This is presumably linked to the fact that this is how we keep score among those who play this game.

The most effective method for resolving authorship disputes is to take steps early on to avoid them. This requires establishing clear expectations and policies for the team, ideally incorporated into the collaboration's foundational documents. At the research-project level, remind participants of these policies at the earliest possible stage of a research or technical development project that could lead to a publication. Even collaborations that have clearly defined policies face gray-area authorship issues that can paralyze progress and generate long-lasting bad feelings. It is common practice in large collaborations to require a declaration at the start of any research project and open up participation to any interested teammates. Even with the best of intentions and efforts, contention over authorship issues can consume considerable time and energy, and can extract an emotional toll on all concerned. Cliques, factions, and alliances can lead to internal tensions. Make sure that newly joined collaboration members understand the team's authorship expectations and policies.

Disagreements about authorship usually focus on two issues: inclusion and the order of people's names. In collaborations that have established a two-tier authorship scheme, where the lead names are those most heavily engaged in generating that publication and others follow (often alphabetically), you can encounter people "below the line" who think their contributions merit inclusion in the first-tier author list. One way to dispel these issues early on is to include the fully ordered author list in early drafts of the paper, so that no one is caught off-guard late in the process. This allows for the resolution of authorship issues to occur in parallel with hammering out the final version, so delays are minimized.

Your first approach to resolving authorship disagreements should be to have a candid and open discussion among the individuals who have differences of opinion. This should be a professional and respectful dialogue, ideally leading to a resolution. If that fails, the parties can engage any dispute-resolution mechanism(s) that exist in the collaboration structure. This might include an ombudsperson or executive committee. If all efforts to reach a compromise solution through these mechanisms fail, you really have no option other than turning to project leadership for assistance in reaching a resolution.

7.5   ESTABLISH EFFECTIVE COMMUNICATION WITH PROJECT MANAGERS

One essential aspect of being successful in a large construction project is building and maintaining a productive, respectful, and fruitful relationship with project managers at all levels. The earlier chapters of this book are meant to help you, as a scientist, better understand and appreciate the perspective, tools, and vocabulary that our project management colleagues bring to the table. You can enhance your relationship with the project management team (and your value to the project) by supporting their aspects of the program. Learn about the processes that are used for identifying new risks and hazards and be a knowledgeable contributor in those areas. Become engaged in discussions around contingency resource allocation, from the perspective of overall project success as opposed to your own subarea. In short, be a person who helps to bridge the divide between scientists and managers on the project.

7.6   MEETINGS, SO MANY MEETINGS! USE THEM EFFECTIVELY

One area where some thought and advance planning can enhance your effectiveness is dealing with the astonishing number of meetings that seem to be an integral part of retaining cohesion in a large enterprise. You have likely already attended meetings that span a spectrum of effectiveness and utility. Whether the meetings are largely decisional, informational, or collaborative/coordinating, or are working sessions to hammer out a solution to some technical or administrative issue, it's worth considering what steps you can take to make them a productive use of people's time. Distributing an agenda in advance keeps the meeting focused. Ending each meeting with a crisp review of action items and deadlines helps to maintain forward momentum. Canceling standing meetings that have no clear purpose that particular week can free up valuable time. Consider information dissemination mechanisms other than face-to-face meetings. Help management consider just who should be attending what gatherings. If you add up the hours per week that team members on large projects spend in meetings, you'll come away motivated to make sure we make the very best use of that time.

If you end up being the person responsible for running one or more of these meetings, then you have a great deal of agency in employing best practices, including:

- Distribute an agenda in advance, even if it's very terse.
- Keep the conversation focused on the agenda items.
- Draw out contributions from all knowledgeable participants.
- Enforce the project's code of conduct. Intervene if individuals fall short of expectations.
- Designate specific individuals as liaisons to other elements of the project structure, with the explicit remit of being a conduit for information flow and exchange.
- Consider alternative communication paths such as Slack channels, wiki pages, and technical memos to replace or supplement information exchange during meeting times.
- Evaluate on a regular basis the cadence, duration, and participant list for standing meetings. Are they all still essential, given the opportunity cost in people's time?
- It sometimes helps to work out political, administrative, fiscal, and/or technical compromise solutions with individuals ahead of time, so that the meeting endorses this solution while avoiding outright conflict.
- End each meeting with a crisp summary of action items, the team member(s) responsible for the work, and deadlines. Taking stock of open action items is a good way to start off subsequent meetings.

7.7 LEAVE A PAPER TRAIL

Large technical projects can take many years to complete. People come and go. Memories are fallible. Not only do people forget things—projects can as well. As you work through issues and problems, document the outcome of those deliberations with an appropriate level of technical detail. If any custom software was developed, link the code repository location to the technical memo. The same goes for datasets.

Generating the technical documentation is the easy part. More difficult is making it easy to find five, ten, or more years down the line. One trick to assist in this is to link your items in the project's document repository to project review materials. Including a list of relevant documents in review presentation overheads can serve as a good breadcrumb. People often try to find project documentation by looking back at review presentations and materials. If your document management system includes keyword searches, pick them judiciously.

ADDITIONAL SUGGESTIONS SPECIFICALLY FOR JUNIOR SCIENTISTS

This chapter makes some additional suggestions for how a junior scientist might make the most of being part of a major project, and some thoughts on navigating some of the corresponding challenges. All the considerations in the previous chapter apply to junior scientists as well, so make sure you read that section as well. As with many of the aspects discussed elsewhere, the issues that arise are often an amplified version of those that arise in the context of smaller-scale endeavors. The comments that follow apply whether your participation is mainly on a construction project or as a member of a large science collaboration.

Being part of large team provides remarkable learning opportunities, both on technical matters and human factors. You can gain expertise in the details of the facility, instrumentation, and/or data processing pipeline. There are opportunities for leadership and room for professional growth.

At the same time, there is a legitimate concern of being swallowed up in a large enterprise with one's contributions unrecognized and underappreciated. This is particularly vexing if paper authorship is alphabetical.

The sections that follow provide some suggestions on how to navigate your way through all this.

8.1 DO YOUR HOMEWORK, THEN TAKE SOME INITIATIVE

When you first start attending team meetings after joining a large team, things are likely going to seem rather unfamiliar. Unless you gained substantial experience in big-team science earlier in your career, this will be an adjustment. There will be more resources, but it never seems like enough. Consequences of failure will be higher, but so will the payoff that comes from success. You'll be seeing new faces and hearing unfamiliar acronyms. You can come up to speed much more rapidly if you invest time in reading the project's documentation: proposals, a project "Science Book" if there is one, the implementation and management of plans, and so on. Don't limit yourself to reading just the science narratives. Get familiar with the budget breakdown and management structure. Either print out or make for yourself some organizational charts that include names of people and their managerial relationships. Pay particular attention to the aspects that surround your area of engagement. Find someone who would be willing, for example, to let you buy them lunch in return for helping you decode both acronyms and understand what the real relationship structures (and tensions) are on the project.

I very much hope that you will have already established a network of mentors earlier in your career. This is a group of people you feel comfortable turning to for advice and support. You should now extend this mentoring network to include people within your new project team.

An important element of learning how the project functions is to identify and understand both the formal and informal communication channels. Do your new project's people exchange information in meetings, through the exchange of documents, through Slack channels or the equivalent, or over meals and coffee breaks? Different parts of the project might have differing dominant communication paths. In the early stages of your project participation, it's best to adapt your communication approach to what exists. Later on, if you think something different might work better, you can certainly advocate for that—but probably not right away.

Once you have the project's relationship and communication structures sketched out, find a manageable upcoming task that comes up in discussion, and volunteer to help out. Perhaps you could assist with a rough calculation, analysis of preliminary data, or prototyping of some aspect of the instrumentation. This does entail some element of risk on your part but stepping up as someone willing to extend a helping hand will help establish your value to your colleagues. Depending on the nature of your relationship with a supervisor, if you have one, you might need to clear this plan with them ahead of time. You might not initially have the perfect skill set match to the things that need doing, but how else will you learn, adapt, and grow?

8.2 BUILD A REPUTATION BY DOING THINGS

Success of a large-team endeavor, whether it's constructing a new facility or carrying out an ambitious collaborative science project, depends on coordinated shared effort and consequent progress. That progress is built upon coordinated individual contributions. It will come as no surprise that not all team members make equally substantive contributions. Some social scientists have posited that this is an example of a more general power law distribution (do a search on "Pareto Principle" if you're interested). You should strive to be one of the individuals who are seen as being central to the project's success. This is established by being someone who delivers the goods. But note the use of the phrase "coordinated shared effort." You can think of each person's contribution as a vector in some abstract space. Uncoordinated contributions don't add up as coherently as when the individual vectors have substantial alignment. That requires communication, coordination, and cooperation; strive to be communicative and collaborative. Your influence and impact on the project can grow more rapidly than you might think. I've seen numerous examples where a motivated, clever, agile, but very junior problem-solver has filled the sails of a project's ship. That breath of fresh air can have a significant positive impact on morale and leads to respect and gratitude across the team. The pitfall here is for that success to inflate your ego.

8.3   CARVE OUT A LEADERSHIP NICHE

As your involvement progresses, see if you can identify an element of the project where you can serve as convener, coordinator, or leader. Sometimes these are elected positions, and sometimes appointed. In the spirit of the discussion above about coordinated efforts being more effective, your initial leadership role will likely amount to coaxing colleagues into making useful and coordinated contributions rather than pursuing their own passion of the moment. In general, scientists don't like to be told what to do. This will be a good opportunity to hone your leadership skills by finding ways to entice collaborative contributions that are aligned with project goals.

Once you've gained some experience with an initial position of leadership and responsibility, you can take stock and see what blend of individual or team science, technology development, project management, teaching and training, and scientific leadership suits you best. Then you can begin to craft an explicit plan for further professional development to meet your objectives.

8.4 MAKE AND EXECUTE A PLAN FOR PROFESSIONAL GROWTH

Being mindful about your goals, near-term and long-term, will help you make a plan for how to achieve them. That plan can then inform your decisions on how you spend your time, which is arguably the most precious resource under your control. It's all too easy to fall into a repetitive cycle of weekly meetings intermingled with crisis response, and subsequently lose track of your own longer-term aspirations. Set aside time on your calendar, perhaps quarterly, to take stock of your development and progress. Do self-assessments. What have you done well? What could have been done better? Are you maturing professionally? What leadership skills come naturally to you (which are all too easy to overlook) and which ones present opportunities for improvement? Have you completed your work on the timetable you expected? Was the last paper you worked on submitted when you expected? What's holding up finishing that internal technical memo?

Having the benefit of the perspective of your teammates on what you're doing well and what could use improvement is very beneficial. However, despite all the project management structure that gets wrapped about any big-team undertaking, individual performance feedback is seldom in my experience a strong point, especially for PhD scientists. That can likely be traced to the academic origins of many of the project's senior participants. Performance evaluation and feedback are not particularly strong points within academia either. That makes it a little awkward to solicit constructive feedback from elements of the organizational chart that are above, below, and alongside you. The best advice I can offer in this regard is to find informal ways to gain this information. Perhaps you could help set up an informal network of junior scientists who are all in the same circumstance? It's likely that the Project Manager will have sympathy for this situation, and might be willing to have a conversation with you on this topic.

It's always difficult to hear criticism, but all of us have areas that are ripe for professional improvement. Be open to the suggestions you hear and fold them into your professional development plan. You might also benefit from identifying individuals on the project whose leadership style you find worthy of emulation and learn from how they handle various situations.

8.5   MAINTAIN EXTERNAL VISIBILITY

One of the challenges you face as a participant in a large-team project is building and/or sustaining visibility and professional relationships with parts of your community that aren't colleagues on the project. Your long-term career and perhaps the prospect of landing your next job may depend on both your internal (within-project) and external (outside-project) reputation. Many factors go into how other people value and rank your contributions: your perceived scientific contributions/leadership, your technical competence/creativity, your abilities as a communicator/teacher, and your track record for nurturing/training/mentoring other junior colleagues.

You will need your contributions, skills, and traits be apparent to people who are outside the project. There are a variety of ways you can approach this. If possible, you could sustain some aspect of your scientific engagement from your life before joining the current project and contribute to follow-on nonproject papers while still meeting your project obligations. You could establish a leadership role producing a project-based science or technical paper, or represent the project at scientific meetings and technical workshops. Many contemporary big-team projects set up an internal management structure to apportion out these "project ambassador" roles across participants, often paying attention to who might be on the job market. Make sure you understand and follow whatever policies might exist for your team.

Attending relevant meetings, regardless of whether you're giving a talk or not, is another way to establish and sustain relationships with people outside of your main project. Go to talks of interest, engage with the community, and strike up conversations during the all-important coffee breaks.

8.6   TAKE THE LEAD ON PAPER AUTHORSHIP

While your CV might contain, depending on authorship policies, a long list of collaboration publications, scientists often highlight the ones in which they played a central role. Try to find opportunities to play a leadership role in developing and submitting a publication to a peer-reviewed journal, be it for a technical/methods paper, a data release, a collaboration science results presentation, or a numerical simulations results description.

Even in large collaborations and projects, usually a handful of people drive this process for each publication. Being one of those people puts you in a position of prominence that others will notice later, when writing letters of recommendation on your behalf.

The management of a publication with many authors is a challenge unto itself, and advance preparation can save you a lot of time later on. If the collaboration you joined is sufficiently mature, this problem will have been at least partially solved. Ask around and see what tools exist to manage version control and to solicit and incorporate feedback. You can also incorporate suggestions from "Ten Simple Rules for Managing Communications with a Large Number of Coauthors" (Muscarella and Poorter 2022), and similar references. Be sure you adhere to collaboration policies on declaring a paper, authorship inclusion, internal reviewing, and publication charges.

8.7   BE A MENTOR TO OTHERS

Once you have a sense of how the project functions and how to be effective, take the initiative to reach out to new or junior members of the team and offer to serve as a mentor. Not only will this provide some personal and professional satisfaction from "paying it forward," it will strengthen your skills in mentoring.

CLOSING THOUGHTS

The scientific ecosystem comprises important research being conducted at a variety of scales, ranging from single individuals to small teams to large collaborative projects. Over their careers, scientists can migrate in both directions, ideally following the most compelling and interesting open questions. Entire subfields can undergo a transformation of scale over the course of a scientist's career. This can lead to sociological tensions and an ongoing examination of values and priorities. Navigating this landscape of research opportunities, making informed choices, and adapting to new collaborative configurations is challenging.

We face vexing questions about the scale-of-science at the national scale and at the individual level. National and international discussions about the optimal scale and complexity of science research involves many factors. Do tabletop projects provide a better training ground for the next generation than having graduate students work only in large collaborative teams? Is there better discovery potential with more numerous smaller projects rather than one big one? Should we sustain the careers of individuals who have gone at risk by working on project(s) that were not selected for full funding? How do we strike the right balance between single-PI grants and investments in major facilities, given fixed budgets? Who should make those decisions?

This continual re-examination of priorities and goals is healthy, but the facts suggest (Marušić, Bošnjak, and Jerončić 2011; Patel et al. 2019; Wuchty, Jones, and Uzzi 2007; Schulman et al. 1997) an overall trend toward science increasingly being done in larger collaborative teams. Across many fields in the past few decades there has been an inexorable consolidation of smaller-scale projects into ever-larger and fewer experiments. Past proliferations of ground-based cosmic microwave background experiments are evolving into a single project, CMB-S4. Smaller dark matter direct detection experiments have consolidated into fewer but larger projects with improved sensitivity. Nuclear physics experimental facilities at universities have been steadily closed down in favor of larger accelerators situated at national laboratories. Synchrotron light sources are increasingly being used by user communities drawn from a wide range of subfields. These evolutions have been largely guided by consensus of the science community, through decadal surveys and similar activities.

While acknowledging this trend toward larger and increasingly expensive experiments and facilities, we must also recognize the mixed track record of success in predicting both cost and schedule for their completion, I believe the science community has a responsibility to step up and contribute as best we can to making these projects as successful as possible. In my view, that requires that scientists be fully engaged in the planning, construction, optimized operation, and scientific exploitation of these facilities.

My hope is that the material presented in this book can help my fellow scientists who are stepping up to these challenges.

APPENDIX 1: REVIEW PRESENTATION CHECKLIST

This list summarizes, as a checklist, the suggestions made in chapter 5 about preparing a presentation for a project review.

- Determine main messages to be conveyed to reviewers, based on review charge
- Find where this talk resides in agenda sequence, and what reviewers will have seen
- Import review slide template and use it
- Use slide titles to convey main message per overhead
- Convey respect and adherence to budget, safety, and project management protocols
- Show how this material relates to project WBS and risk elements
- Anticipate reviewer questions and incorporate back-up slides as appropriate
- Include back-up slides for a diversity of reviewer expertise
- Divide talk into main presentation and back-up sections
- Rehearse—does a no-interruptions delivery take about half the allotted time?
- Coordinate with teammates on answering reviewer questions on specific topics
- Ensure that talk transfers over to presentation computer with no font corruption

APPENDIX 2: EXAMPLE REVIEW CHARGE

This appendix contains the charge to the review panel that was convened for a multiagency (NSF and DOE) annual progress review for the Vera C. Rubin Observatory (called the Large Synoptic Survey Telescope at the time).

Charge for the Large Synoptic Survey Telescope Annual Progress Review
August 27–30, 2019, Tucson, AZ

The National Science Foundation (NSF) Division of Astronomical Sciences (AST) and the Department of Energy (DOE) Office of High Energy Physics (HEP) will jointly conduct a Progress Review of the Large Synoptic Survey Telescope (LSST) Project from August 27 to 30, 2019, in Tucson, Arizona. The purpose of this review is to conduct the annual assessment of the construction project status and future plans as well as an assessment of the planning and execution of the activities during the commissioning phase. The review will be conducted to conform to both NSF and DOE requirements for annual progress reviews. In parallel, there will be a separate Earned Value Management System (EVMS) surveillance review covering only the NSF-funded activities (see appendix).

Construction of the LSST Project is jointly supported by NSF and DOE. The NSF is the lead agency, funding construction of the telescope, site facilities, data infrastructure, systems engineering, and education and public outreach capabilities, with an award from the Major Research Equipment and Facilities Construction (MREFC) program restricted by the National Science Board not to exceed $473 million. The DOE-deliverable portion of the LSST Project is the imaging camera system with a total project cost of $168 million funded as a Major Item of Equipment project. With private support of $39 million, the full construction cost is $680 million.

Commissioning preparations and activities began in fiscal year (FY) 2018 and are being planned and executed by the construction project team. Support for commissioning phase activities is included in the MREFC project funded by NSF. The corresponding activities on the DOE side are supported by an operations funding line as they are not included in the Camera MIE project.

The LSST facility is scheduled to begin its full 10-year survey operations on October 1, 2022. Ramp-up of pre-operations activity began in FY 2019; the pre-operations activities and funding are part of a separate planning exercise.

The LSST Project is organized by Work Breakdown Structure (WBS) under the headings "1. Project Management Office," "2. Data Management," "3. Camera," "4. Telescope and Site," "5. Education and Public Outreach," and "6. Systems Engineering and Commissioning." The Project is managed as a unified, single project across NSF- and DOE-funded scope. It has a complex organization, in part because it is supported by two separate federal agencies with different policies and procedures. The panel is asked to consider throughout this technical, cost, and schedule review, how well the integrated project management is working, and to draw attention to any interface issues they may identify.

The LSST Project will share all necessary documentation with the review panel through a Web-based repository, available at least one week before the review starts. This will include the Project Execution Plan, which contains the performance baseline against which success is to be measured. We expect that pre-review communications with the panel by both the LSST Project and the Agencies will be by email.

Edward Ajhar is the NSF program manager for LSST and will serve as the NSF AST contact for the review. Helmut Marsiske is the DOE program manager for the LSST Camera and will serve as the HEP contact for the review. Kathy Turner is the DOE program manager for LSST Camera commissioning and operations.

The review panel is requested to prepare a closeout presentation for the project, containing their major recommendations, which will be given on the last day. The full written report should be submitted to both agencies within 60 days of the review.

CHARGE TO THE PANEL FOR THE PROGRESS REVIEW

Considering the LSST project's performance to date and the execution plan for the future, including technical scope, cost, schedule, and the safety and risk management plans, can the facility be constructed and commissioned as planned, and is the project effectively coordinating its activities with the pre-operations team?

The panel should answer the following questions, as noted under primary numbering and in bold-face. Wherever possible, any identified shortcomings should be accompanied by recommendations that the panel believes will correct the problem. Although there are specific questions in this charge, the panel is asked to examine project activities broadly and to draw attention to any issue they should happen to notice, even if it does not appear in this charge.

Is the LSST project progressing as planned, and are they making appropriate plans for future construction and commissioning work?

Are all activities consistent with the baseline project objectives as described in the Project Execution Plan? Is there adequate progress and planning across all Work Breakdown Structure elements, including both in-house efforts and external procurements and contracts? Are there appropriate plans for realizing opportunities and for mitigating risks? Are the plans for assembly, integration, test, and commissioning phases complete, achievable, and well understood by the commissioning team?

Are the current cost and schedule (C&S) performance and their future trends acceptable?

Are the budgets and contingencies for the full scope of the NSF deliverables and DOE deliverables covered by each agency's funding plans? Do the performance to-date and C&S trends give confidence that the project can complete successfully on time and within budget? Is there adequate reporting for both funding agencies? Have changes to the Project Management Control System (PMCS) been properly incorporated? Is the change control process solid and is it being followed correctly?

Is the project management functioning well?

Are there clear lines of authority and responsibility? Are the interfaces between DOE- and NSF-supported activities managed appropriately? Are new risks being uncovered and are identified risks being actively mitigated and/or retired? Are there concerns over the distributed nature of the second level WBS teams and how well they are communicating and working together? Does the systems engineering team monitor progress and performance across the project using proper methodology? Is the systems engineering documentation being suitably created and refined as the project progresses? Are all necessary written procedures in place, documented, and followed? Are contracts and procurements properly monitored? Is the project staffing adequate for the work?

Is the planning for the transition to operations adequate for this stage of the project?

Will essential materials, including manuals, maintenance plans, test reports, and as-built drawings be ready when needed? Is planning for the transition of personnel sufficiently developed and being appropriately communicated to staff?

Are the Education and Public Outreach (EPO) activities properly planned and executed?

Is the EPO team the right size with the right skills? Is it engaged and integrated as appropriate?

Are Environment, Safety & Health (ES&H) issues managed appropriately?
Does the project have an acceptable safety record? Are Integrated Safety Management Principles being followed?

Has the project responded satisfactorily to recommendations from previous reviews?

NSF is conducting a separate but parallel surveillance review of the project's EVM System used for the NSF-funded scope of the project, excluding the DOE-supported items. (EVM Systems at the DOE labs undergo separate reviews). This review ensures that the NSF-accepted EVMS is being maintained, and continues to provide reliable project performance data.

The EVMS Surveillance is the process of reviewing the implementation and use of the accepted EVMS processes and procedures for the project. The 2019 Surveillance Review seeks to accomplish three goals:

- Confirm implementation of EVMS changes in response to the August 2018 review and assess effectiveness.
- Review LSST Project Team's adherence to their EVMS processes and procedures in accordance with the LSST Project Controls documentation.
- Review the timeliness, accuracy, and reliability of project performance data provided by LSST.

All process categories of EVM will be considered during this system surveillance with a focus on two of the five process categories. This EVMS surveillance will be based upon the remaining work and content that is specific to the guidelines being reviewed. The surveillance will address the content of the LSST Project Controls System Description with focus on changes from the August 2018 review and the accuracy of the EVM system over the last three months.

The outcome will be a written report to be submitted to the NSF Program Officer by October 31, 2019. Although NSF-focused, this report will be shared with DOE.

NOTES

CHAPTER 2

1. We distinguish here between risks, which are related to technical failure or shortcomings, and hazards that are personnel safety concerns.

CHAPTER 3

1. The author still suffers from carpal tunnel syndrome that resulted from an early-career forty-eight-hour proposal preparation marathon that was a consequence of inadequate margin. And no, it wasn't funded. Coincidence? Probably not . . .

2. Beware—this review is sometimes called a "CDR," which is a term also used for a Critical Design Review, which is a totally different beast that occurs much later in the project life cycle.

3. Research & Innovation

4. European Institute of Innovation & Technology.

5. Sustainable Development Goals.

6. Small and Medium Enterprises.

7. Knowledge and Innovation Communities.

CHAPTER 4

1. In-kind contributions are when participants contribute labor or pieces of hardware, rather than sending the project a check.

2. From November 19, 2020.

3. Some firms negotiate what are called cost-plus contracts with the government. Under this scheme, their costs are reimbursed, along with a fee that gives a guaranteed modest profit.

BIBLIOGRAPHY

"About CERN—CERN Environment Report 2019–2020." HSE Unit at CERN. 2020. https://hse.cern/environment-report-2019-2020/about-cern.

Adelberger, E. G., B. R. Heckel, C. W. Stubbs, and W. F. Rogers. 1991. "Searches for New Macroscopic Forces." *Annual Review of Nuclear and Particle Science* 41, no. 1: 269–320.

Alcock, C., R. A. Allsman, D. R. Alves, T. S. Axelrod, A. C. Becker, D. P. Bennett, K. H. Cook, et al. 2000. "The MACHO Project: Microlensing Results from 5.7 Years of Large Magellanic Cloud Observations." *Astrophysical Journal* 542, no. 1: 281.

"Atacama Large Millimeter Array." 2022. Wikipedia. https://en.wikipedia.org/w/index.php?title=Atacama_Large_Millimeter_Array&oldid=1124443963.

Billings, Lee. 2010. "The Telescope That Ate Astronomy." *Nature* 467, no. 7319: 1028–1030.

Boehm, Barry. 1986. "A Spiral Model of Software Development and Enhancement." *ACM SIGSOFT Software Engineering Notes* 11, no. 4: 14–24.

Boehm, Barry W. 1988. "A Spiral Model of Software Development and Enhancement." *Computer* 21, no. 5: 61–72.

D'Ippolito, Beatrice, and Charles-Clemens Rüling. 2019. "Research Collaboration in Large Scale Research Infrastructures: Collaboration Types and Policy Implications." *Research Policy* 48, no. 5: 1282–1296.

"ESA Science & Technology—Cosmic Vision." n.d. Accessed January 15, 2023. https://sci.esa.int/web/cosmic-vision/home.

Flyvbjerg, Bent, Nils Bruzelius, and Werner Rothengatter. 2003. *Megaprojects and Risk: An Anatomy of Ambition*. Cambridge: Cambridge University Press.

Frank, Melvin. 2018. DOE G 413.3-9A (2018), "Project Reviews for Capital Asset Projects." https://www.directives.doe.gov/directives-documents/400-series/0413.3-EGuide-17-admchg1.

"HET-HEPAP.Pdf." 2016. Google Docs. https://drive.google.com/file/d/0B3IXprj4oX2sU0VvczR6VzdheDQ.

"Horizon Europe Strategic Plan." n.d. Accessed January 15, 2023. https://research-and-innovation.ec.europa.eu/funding/funding-opportunities/funding-programmes-and-open-calls/horizon-europe_en. See also European Commission, Directorate-General for Research and Innovation. 2021. *Horizon Europe: Strategic Plan 2021–2024*. Publications Office of the European Union. https://data.europa.eu/doi/10.2777/083753.

Ilakovac, Vesna, Kristina Fister, Matko Marusic, and Ana Marusic. 2007. "Reliability of Disclosure Forms of Authors' Contributions." *Cmaj* 176, no. 1: 41–46.

InterAcademy Partnership. 2016. *Doing Global Science: A Guide to Responsible Conduct in the Global Research Enterprise*. Princeton, NJ: Princeton University Press.

Ivezić, Željko, Steven M. Kahn, J. Anthony Tyson, Bob Abel, Emily Acosta, Robyn Allsman, David Alonso, et al. 2019. "LSST: From Science Drivers to Reference Design and Anticipated Data Products." *Astrophysical Journal* 873, no. 2: 111.

Kirk, William. 1970. "An Informal History of SLAC." https://www.slac.stanford.edu/history/history.shtml.

Kupferschmidt, Kai. 2019. "European Commission Kills Billion-Euro Flagship Concept." *Science* 364 (6441): 618–619. https://doi.org/10.1126/science.364.6441.618.

LCLS project. 2009. "LCLS WBS Dictionary." https://www-ssrl.slac.stanford.edu/lcls/doe_reviews/2009_may_13-15_lcls/Documents/LCLS_WBS_Dictionary_LCLB_r4.pdf.

Mankins, John C. 2009. "Technology Readiness Assessments: A Retrospective." *Acta Astronautica* 65, nos. 9–10: 1216–1223.

Martin, Joseph D. 2018. *Solid State Insurrection: How the Science of Substance Made American Physics Matter*. Pittsburgh: University of Pittsburgh Press.

Marušić, Ana, Lana Bošnjak, and Ana Jerončić. 2011. "A Systematic Review of Research on the Meaning, Ethics and Practices of Authorship across Scholarly Disciplines." *Plos One* 6, no. 9: e23477.

McNutt, Marcia K., Monica Bradford, Jeffrey M. Drazen, Brooks Hanson, Bob Howard, Kathleen Hall Jamieson, Véronique Kiermer, et al. 2018. "Transparency in Authors' Contributions and Responsibilities to Promote Integrity in Scientific Publication." *Proceedings of the National Academy of Sciences* 115 (11): 2557–2560.

"MRI Frequently Asked Questions (Nsf15012) | NSF—National Science Foundation." n.d. Accessed January 15, 2023. https://www.nsf.gov/pubs/2015/nsf15012/nsf15012.jsp.

Munroe, Randall. n.d. "JWST Delays." xkcd. Accessed January 15, 2023. https://xkcd.com/2014/.

Munroe, Randall. n.d. "Software Development." xkcd. Accessed January 15, 2023. https://xkcd.com/2021/.

Muscarella, Robert, and Lourens Poorter. 2022. "Ten Simple Rules for Managing Communications with a Large Number of Coauthors." *PLOS Computational Biology* 18 (6): e1010185.

Naslund, Dag, and Rahul Kale. 2020. "Is Agile the Latest Management Fad? A Review of Success Factors of Agile Transformations." *International Journal of Quality and Service Sciences* 12, no. 4: 489–504.

National Academy of Sciences. 2009. *On Being a Scientist: A Guide to Responsible Conduct in Research*. 3rd ed. Washington, DC: National Academies Press.

Norwich, John Julius. 2003. *A History of Venice*. London: Penguin UK.

"NPR 7123.1C—Chapter1." n.d. Accessed December 22, 2022. https://nodis3.gsfc.nasa.gov/displayDir.cfm?Internal_ID=N_PR_7123_001C_&page_name=Chapter1&search_term=NASA%20Systems%20Engineering%20Processes%20and%20Requirement.

Olson, Maynard V. 1993. "The Human Genome Project." *Proceedings of the National Academy of Sciences* 90, no. 10: 4338–4344.

Patel, Vanash M., Pietro Panzarasa, Hutan Ashrafian, Tim S. Evans, Ali Kirresh, Nick Sevdalis, Ara Darzi, and Thanos Athanasiou. 2019. "Collaborative Patterns, Authorship Practices and Scientific Success in Biomedical Research: A Network Analysis." *Journal of the Royal Society of Medicine* 112, no. 6: 245–257.

Ray, Debraj, and Arthur Robson. 2018. "Certified Random: A New Order for Coauthorship." *American Economic Review* 108, no. 2: 489–520.

Riess, Adam G., Alexei V. Filippenko, Peter Challis, Alejandro Clocchiatti, Alan Diercks, Peter M. Garnavich, Ron L. Gilliland, et al. 1998. "Observational Evidence from Supernovae for an Accelerating Universe and a Cosmological Constant." *The Astronomical Journal* 116, no. 3: 1009.

Riordan, Michael, Lillian Hoddeson, and Adrienne W. Kolb. 2015. *Tunnel Visions: The Rise and Fall of the Superconducting Super Collider*. Chicago: University of Chicago Press.

Ritz, Steve, Hiroaki Aihara, Martin Breidenbach, Bob Cousins, Andre de Gouvea, Marcel Demarteau, Scott Dodelson, et al. 2014. "Building for Discovery: Strategic Plan for US Particle Physics in the Global Context." Report of the Particle Physics Project Prioritization Panel (P5), HEPAP Subcommittee. https://science.osti.gov/-/media/hep/hepap/pdf/May-2014/P5MayHEPAP-Ritz.pdf.

Samuelson, Pamela. 2020. "AI Authorship?" *Communications of the ACM* 63, no. 7: 20–22.

Schulman, Eric, James C. French, Allison L. Powell, Guenther Eichhorn, Michael J. Kurtz, and Stephen S. Murray. 1997. "Trends in Astronomical Publication between 1975 and 1996." *Publications of the Astronomical Society of the Pacific* 109, no. 741: 1278.

Stubbs, Christopher. 1997. "The Serious Business of Listing Authors." *Nature* 388, no. 6640: 320.

Terrell, Stefanie M. 2018. NASA Work Breakdown Structure (WBS) Handbook. No. NASA/SP-2016-3404/REV1.

Transformer, G. P., Jr., End X. Note, M. S. Spellchecker, and Roman V. Yampolskiy. 2020. "When Should Co-authorship Be Given to AI?" https://philarchive.org/archive/GPTWSCv1.

U.S. DOE. 2020. "Earned Value Management System (EVMS) and Project Analysis Standard Operating Procedure (EASOP)." DOE-PM-SOP-05-2020.

U.S. DOT. 1998. "Office of Inspector General Audit Report Advance Automation System." U.S. Dept of Transportation.

U.S. GAO. 2021. "James Webb Space Telescope Project Nearing Completion, but Work to Resolve Challenges Continues." U.S. Government Accountability Office. https://www.gao.gov/products/gao-21-406.

"US NSF—MPS—AST—Division Portfolio Review." 2012. https://www.nsf.gov/mps/ast/ast_portfolio_review.jsp.

Wallace, Nicholas. 2020. "European Research Budget Gets Unexpected €4 Billion Boost." 2020. https://www.science.org/content/article/european-research-budget-gets-unexpected-4-billion-boost.

Willman, Paul. 2014. *Understanding Management: The Social Science Foundations*. Oxford: Oxford University Press.

Wuchty, Stefan, Benjamin F. Jones, and Brian Uzzi. 2007. "The Increasing Dominance of Teams in Production of Knowledge." *Science* 316, no. 5827: 1036–1039.

Young, Chuck, Orice W. Brown, and James-Christian Blockwood. 2020. "Cost Estimating and Assessment Guide: Best Practices for Developing and Managing Program Costs." Washington, DC: Government Accountability Office.

Zastrow, Mark. 2020. "Arecibo Down, but Not Necessarily Out, Following Cable Failure." Astronomy.com. 2020. https://astronomy.com/news/2020/08/arecibo-down-but-not-necessarily-out-following-cable-failure.

Page numbers followed by f refer to figures; page numbers followed by t indicate tables.